Sponsored by the
European Association of Neurosurgical Societies

Advances and Technical Standards in Neurosurgery

Edited by

L. Symon, London (Editor-in-Chief)
J. Brihaye, Brussels
B. Guidetti, Rome
F. Loew, Homburg/Saar
J. D. Miller, Edinburgh
H. Nornes, Oslo
E. Pásztor, Budapest
B. Pertuiset, Paris
M. G. Yaşargil, Zurich

Volume 14

Springer-Verlag
Wien   New York   1986

With 71 partly colored Figures

Product Liability: The publisher can give no guarantee for information about drug dosage and application thereof contained in this book. In every individual case the respective user must check its accuracy by consulting other pharmaceutical literature.

This work is subject to copyright

All rights are reserved, whether the whole or part of the material is concerned, specifically those of translation, reprinting, re-use of illustrations, broadcasting, reproduction by photocopying machine or similar means, and storage in data banks

© 1986 by Springer-Verlag/Wien

Softcover reprint of the hardcover 1st edition 1986

Library of Congress Catalog Card Number 74-10499

ISSN 0095-4829

ISBN-13: 978-3-7091-7466-1   e-ISBN-13: 978-3-7091-6995-7
DOI: 10.1007/978-3-7091-6995-7

Preface

As an addition to the European postgraduate training system for young neurosurgeons we began to publish in 1974 this series devoted to Advances and Technical Standards in Neurosurgery which was later sponsored by the European Association of Neurosurgical Societies.

The fact that the English language is well on the way to becoming the international medium at European scientific conferences is a great asset in terms of mutual understanding. Therefore we have decided to publish all contributions in English, regardless of the native language of the authors.

All contributions are submitted to the entire editorial board before publication of any volume.

Our series is not intended to compete with the publications of original scientific papers in other neurosurgical journals. Our intention is, rather, to present fields of neurosurgery and related areas in which important recent advances have been made. The contributions are written by specialists in the given fields and constitute the first part of each volume.

In the second part of each volume, we publish detailed descriptions of standard operative procedures, furnished by experienced clinicians; in these articles the authors describe the techniques they employ and explain the advantages, difficulties and risks involved in the various procedures. This part is intended primarily to assist young neurosurgeons in their postgraduate training. However, we are convinced that it will also be useful to experienced, fully trained neurosurgeons.

The descriptions of standard operative procedures are a novel feature of our series. We intend that this section should make available the findings of European neurosurgeons, published perhaps in less familiar languages, to neurosurgeons beyond the boundaries of the authors countries and of Europe. We will however from time to time bring to the notice of our European colleagues, operative procedures from colleagues in the United States and Japan, who have developed techniques which may now be regarded as standard. Our aim throughout is to promote contacts among neurosurgeons in Europe and throughout the world neurosurgical community in general.

We hope therefore that surgeons not only in Europe, but throughout the world will profit by this series of Advances and Technical Standards in Neurosurgery.

The Editors

Contents

Listed in Index Medicus

List of Contributors XIII

A. Advances

Endoneurosurgery: Endoscopic Intracranial Surgery. By Huw B. Griffith, Department of Neurological Surgery, Frenchay Hospital and University of Bristol (U.K.) 3
Historical Introduction 3
The Hopkins Advances 5
Light Sources 7
    Light Concentration 8
    Colour Balance 8
    Heat Output 9
Manipulations 10
Endoneurosurgery and Hydrocephalus 11
Intraventricular Tumour Biopsy 15
Endoscopic Photography 16
Other Diseases 17
Small Flexible Endoscopes 19
Sterilising Techniques 20
Future Developments 21
Artificial Cerebrospinal Fluid Solution 22
References 23

Evoked Potential Monitoring in Neurosurgical Practice. By L. Symon, F. Momma, K. Schwerdtfeger, P. Bentivoglio, I. E. Costa e Silva, and A. Wang, Gough-Cooper Department of Neurological Surgery, Institute of Neurology, London (U.K.) and Department of Neurosurgery, Saarland University Medical School, Homburg/Saar (Federal Republic of Germany) 25
1. The Use of Somatosensory Evoked Responses 26
    Somatosensory Evoked Responses in Vascular Diseases of the Brain 26
    Methods 28

Electrodes and Montage 28
Stimulation 29
Recording Apparatus 29
The Measurement of SSEP a Prognostic Guide in Subarachnoid Hemorrhage 30
The Relationship Between Conduction Time, Prolongation and Clinical Deterioration 34
The Use of Somatosensory Evoked Responses as a Monitor During Intracranial Aneurysm Surgery 37
Techniques 37
SSEP Recording as a Monitor During Temporary Vascular Occlusion .. 40
Somatosensory Evoked Potential Recording and the Use of Somatosensory Evoked Responses in Head Injury Management 43
Clinical Factors Related to Outcome 45
SEP Related to Outcome 47
Progressive Change in SEPs 49

2. Visual Evoked Response Monitoring 51

3. Methodology 52
Recording Technique and Wave Identification 54
VEP as a Guide to Manipulation of the Optic Pathways During Surgery ... 55
Conclusions 57

4. The Use of Brain Stem and Other Auditory Evoked Potentials 58
Brain Stem Auditory Evoked Potentials in Head Injury 58
Use of Brain Stem Evoked Responses During Posterior Fossa Surgery .. 59
Electrocochleographic Monitoring During Acoustic Neuroma Surgery . 61
The Technique of Peroperative Electrocochleography 61

5. Spinal Cord Monitoring 64

References 65

The Biological Role of Hypothalamic Hypophysiotropic Neuropeptides. By K. VON WERDER, Medizinische Klinik Innenstadt, University of Munich, Munich (Federal Republic of Germany) 71

1. Introduction 73

2. Thyrotropin Releasing Hormone (TRH) 75
   2.1. Distribution of TRH 75
   2.2. TRH as a Hypophysiotropic Hormone 77
   2.3. Pathophysiology of TRH 78
   2.4. Clinical Utilization of TRH 78
       2.4.1. Diagnosis of Thyroid Disorders 78
       2.4.2. TRH in the Diagnosis of Hypothalamic-Pituitary Disease 80
       2.4.3. Diagnostic Use of TRH in Disorders of PRL Secretion 82

2.4.4. TRH as a Diagnostic Aid in Acromegaly	83
2.4.5. Therapeutic Aspects of TRH	84
3. Gonadotropin Releasing Hormone (GnRH)	84
3.1. Distribution of GnRH	85
3.2. GnRH as a Hypophysiotropic Hormone	85
3.2.1. Physiological Role of Endogenous GnRH	85
3.2.2. Stimulation of Gonadotropin Secretion with GnRH	86
3.3. Pathophysiology of GnRH Secretion	88
3.4. Clinical Utilization of GnRH	89
3.4.1. GnRH in the Diagnosis of Gonadal Disorders	89
3.4.2. GnRH-Test in Hyperprolactinemic Disorders	90
3.4.3. GnRH in Acromegaly	90
3.4.4. Treatment with GnRH	90
3.5. Superactive GnRH Agonists	92
4. Corticotropin Releasing Hormone (CRH)	94
4.1. Structure of CRH	94
4.2. Distribution of CRH	95
4.3. Measurement of CRH	95
4.4. CRH as a Hypophysiotropic Hormone	96
4.5. Pathophysiology of CRH	97
4.6. Clinical Utilization of CRH	98
4.6.1. Use of CRH in the Differential Diagnosis of Adrenal Failure	98
4.6.2. Use of CRH as a Diagnostic Aid in Cushing's Syndrome	100
5. Growth Hormone Releasing Hormone (GRH)	101
5.1. Structure and Biological Activity of GRH	101
5.2. Distribution of GRH	103
5.3. Measurement of GRH	103
5.4. Clinical Utilization of GRH	105
5.4.1. Biological Activity of GRH in Normal Subjects	105
5.4.2. Evaluation of Anterior Pituitary Function with GRH	107
5.4.3. Use of GRH in the Differential Diagnosis and Treatment of Pituitary Dwarfism	109
5.4.4. GRH as a Diagnostic Aid in Acromegaly	109
6. Somatostatin (SRIF)	112
6.1. Distribution of Somatostatin	113
6.2. Physiological Role of Somatostatin	114
6.2.1. Somatostatin as a Hypophysiotropic Hormone	114
6.2.2. Measurement of Somatostatin	114
6.3. Pathophysiology of Somatostatin	115
6.4. Clinical Use of Somatostatin	115
7. Hypothalamic Hormones Regulating Prolactin Secretion	117
7.1. Prolactin Inhibiting Factor	117
7.2. Prolactin Releasing Factor	118

7.3. Clinical Utilization of Hypothalamic Hormones Regulating PRL Secretion 119

8. Summary 119

References 122

B. Technical Standards

Sphenoidal Ridge Meningioma. By D. FOHANNO and A. BITAR, Clinique neurochirurgicale, CHU Pitié-Salpêtrière, Paris (France) 137

Introduction 137
First Symptoms 139
Clinical Examination 141
Investigations 142
    Plain X-Rays 142
    CT-Scan 143
    EEG 146
    Gamma-Scan 146
    MRI 147
    Angiography 148
General Surgical Considerations 151
The Decision for Operation in Sphenoidal Ridge Meningiomas 155
Preoperative Care 156
Anaesthesia 157
Positioning the Patient on the Operation Table 158
Surgical Management of Sphenoidal Ridge Meningiomas 158
    Management of Deep-seated (Medial) Sphenoidal Meningiomas 159
    Removal of Lateral Sphenoidal Wing Meningiomas 168
    Management of "en plaque" and Invasive Meningiomas 170
Postoperative Care 171
Results 171
Acknowledgments 173
References 173

Congenital Spinal Cord Tumors in Children. By H. J. HOFFMAN, R. W. GRIEBEL, and E. B. HENDRICK, Division of Neurosurgery, University of Toronto and The Hospital for Sick Children, Toronto, Ontario (Canada)   175

1. Introduction: Comments on Classification and Embryology 176
2. Dermoid and Epidermoid Cysts 177
    A. Embryology 177
    B. Clinical Features 178

C. Radiological Investigations	182
D. Surgical Management and Follow-up	183
E. Discussion	185
3. Neurenteric Cysts	185
A. Embryology	185
B. Clinical Features and Presentation	186
C. Radiology	187
D. Surgical Management and Follow-up	187
E. Discussion	188
4. Teratomatous Cysts	188
A. Embryology	188
B. Clinical Features	189
C. Case Presentation	189
5. Intraspinal Teratomas	191
A. Embryology	191
B. Clinical Features	191
C. Case Presentations	191
6. Lipomas	192
A. Embryology	192
B. Clinical Features	193
C. Case Presentations	193
7. Spinal Arachnoid Cysts	195
8. Congenital Malignant Tumors of the Spine	196
9. Summary	197
References	197

Controversial Views of Editorial Board on the Intraoperative Management of Ruptured Saccular Aneurysms ... 201

Introduction ... 201
1. What Is the Place of Lumbar CSF Drainage in the Operative Management of Aneurysms? ... 201
2. In the Presence of Considerable Dural Tension and no Evidence of Hydrocephalus on the CT, what Should Be Done? Should the Dura Be Opened? ... 202
3. If Brain Swelling Is Present Should Brain Resection Be Performed? ... 203
4. In the Presence of an Aneurysm Hematoma Should It Be Removed Before Tackling the Aneurysm Neck? ... 203
5. Is Operating on Aneurysms at the Present Time Justifiable Without a Microscope? What About Loupes? ... 204
6. Is Self Retaining Retraction Absolutely Necessary? ... 204

7. Should Mean Arterial Pressure Be Recorded and Should Measurements of Pulmonary Artery Pressure Be Made?.. 204
8. What Is the Place of Arterial Hypotension and Temporary Clipping?. 205
9. Should Bipolar Coagulation Be Used to Reduce the Size of the Sac or to Reduce the Caliber of Its Neck?.. 206
10. What Is the Preferred Method of Occlusion of the Neck, Which Clip? Are Ligatures ever Necessary?... 207
11. Is Coating a Safe Procedure.. 207
12. What Should Be Done when an Intraoperative Rupture Occurs?........ 208
13. What Should Be Done when there is Obvious Spasm of the Internal Carotid Artery at Operation?.. 208
14. What Is Your View of Cardiac Arrest and Hypothermia?................. 209
15. What Should Be Done when a Junior Surgeon Calls for Help During an Operation... 209
Conclusion 211

Author Index 213
Subject Index 225

List of Contributors

Bentivoglio, P., MD, FRACS, Department of Neurological Surgery, Institute of Neurology, The National Hospital, Queen Square, London WC1N 3BG, U.K.

Bitar, Dr. A., Clinique Neuro-chirurgicale Universitaire, C.H.U. Pitié-Salpêtrière, 83 Boulevard de l'Hôpital, F-75651 Paris Cedex 13, France.

Costa e Silva, I. E., MD, M. Phil., Department of Neurological Surgery, Institute of Neurology, The National Hospital, Queen Square, London WC1N 3BG, U.K.

Fohanno, Dr. D., Professeur agrégé, Clinique Neuro-chirurgicale Universitaire, C.H.U. Pitié-Salpêtrière, 83 Boulevard de l'Hôpital, F-75651 Paris Cedex 13, France.

Griebel, Dr. R. W., Department of Clinical Neurological Sciences, University of Saskatchewan, University Hospital, Saskatoon, Saskatchewan, Canada S7N 0X0.

Griffith, H. B., MA, BM, FRCP, FRCS, Department of Neurosurgery, Frenchay Hospital, Bristol BS16 1LE, U.K.

Hendrick, E. B., MD, BSc (Med), FRCS (C), Division of Neurosurgery, University of Toronto, The Hospital for Sick Children, 555 University Avenue, Toronto, Ontario, Canada M5G 1X8.

Hoffman, H. J., MD, BSc (Med), FRCS (C), FACS, Division of Neurosurgery, University of Toronto, The Hospital for Sick Children, 555 University Avenue, Toronto, Ontario, Canada M5G 1X8.

Momma, F., MD, Department of Neurological Surgery, Institute of Neurology, The National Hospital, Queen Square, London WC1N 3BG, U.K.

Schwerdtfeger, Dr. K., Neurochirurgische Universitätsklinik, D-6650 Homburg/Saar, Federal Republic of Germany.

Symon, TD, FRCS, Department of Neurological Surgery, Institute of Neurology, The National Hospital, Queen Square, London WC1N 3BG, U.K.

Wang, A., MD, PhD, Department of Neurological Surgery, Institute of Neurology, The National Hospital, Queen Sqare, London WC1N 3BG, U.K.

Werder, Professor Dr. K. von, Medizinische Klinik Innenstadt der Universität, Ziemssenstrasse 1, D-8000 München 2, Federal Republic of Germany.

A. Advances

Endoneurosurgery: Endoscopic Intracranial Surgery

HUW B. GRIFFITH

Department of Neurological Surgery, Frenchay Hospital and University of Bristol (U.K.)

With 8 Figures

Contents

Historical Introduction	3
The Hopkins Advances	5
Light Sources	7
Light Concentration	8
Colour Balance	8
Heat Output	9
Manipulations	10
Endoneurosurgery and Hydrocephalus	11
Intraventricular Tumour Biopsy	15
Endoscopic Photography	16
Other Diseases	17
Small Flexible Endoscopes	19
Sterilising Techniques	20
Future Developments	21
Artificial Cerebrospinal Fluid Solution	22
References	23

Historical Introduction

Seventy five years ago the cerebral ventricles of a hydrocephalic child were first inspected *in vivo* by means of an endoscope. The pioneering surgeon was Espinasse of Chicago. The instrument was a cystoscope designed for looking into the adult bladder. This utilised the system of lenses designed by Nitze of Vienna in 1887 illuminated by a distally placed incandescent bulb. This appears to have been applied to a cystoscope by

Newman of Glasgow only three years after Edison had devised the incandescent lamp in 1880. Espinasse's adventure was described by Davis (L'Espinasse 1943) in the following terms:

"The first attempt to remove a choroid plexus for hydrocephalus with which I am familiar was made by Dr. V. L'Espinasse of Chicago. In 1910, L'Espinasse introduced a small cystoscope into the ventricle and fulgurated the plexus bilaterally in 2 infants. One of his patients died post operatively and the second lived 5 years. The method was presented before a local medical society and was not otherwise recorded."

There then followed over the next three decades sporadic developments of interest in the possibilities of this technique. For instance, in 1923 Grant and Fay recorded the taking of black and white photographs of the internal features of the cerebral ventricle of a hydrocephalic child, using a photographic exposure of 40 seconds. Putnam in 1934 made a further advance by describing a new endoscope which consisted of a solid glass rod some 7 mm in diameter, grooved on the side for the electric light carrier and equipped with bipolar metal electrodes ending just in front of the polished flat end of the rod which formed the collecting field lens. Two years later Scarff introduced a "fore-oblique" telescope of conventional design, carrying a movable electrode, moving in a sheath, all conforming fairly closely to the pattern of the cystoscope of that era. Scarff persisted with the technique and by the time he reported his final experience in 1970 had operated upon 19 cases of communicating hydrocephalus (see later) and reported a success rate of approximately 80% as far as control of the hydrocephalus was concerned. However, the documentation of Scarff's cases was not good enough to allow a rigorous assessment of his results. The procedure is described as lasting for some 3 h, and required an opening in the skull approximately 2.5 cm in diameter for manipulation of the endoscope. Although the size of the instrument is not given, from indications such as this it looks as if its calibre was considerable. Evidently there was post-operative trouble with cerebrospinal fluid leakage from the scalp since Scarff recommended an elaborate scalp closure designed to circumvent this complication.

The next technical advance in endoscopy of the brain was reported by Guiot *et al.* in 1963. They used for the first time an endoscope with the field of view illuminated by a powerful external light guided through the endoscopic sheath beside the telescope by a quartz rod. This functioned as a light guide because of its property of total internal reflection. It resulted in a greatly improved view despite the drawbacks of the original Nitze design of telescope. Guiot and his colleagues were able to carry out the operation of third ventriculostomy with this instrument and were successful in making a colour film of the inside of the ventricles. However the sheath was 9.1 mm in external diameter, still too large. Eventually the technique dropped out of use. By the 1960s however, the incongruent glass-fibre bundle for use as a light guide became available and Scarff improved his ventriculoscope in 1963 by substituting what has subsequently become known as "fibre-lighting" for the distal illumination by incandescent bulb. At about the same time the Machida Optical Company of Japan (Ogata *et al.* 1965), using a light guide of 10 000 incongruent fibres each 18 μm in diameter, were able to construct a small endoscope of convential design, the telescope diameter being 3.1 mm, introduced through a trocar of diameter 3.6 mm. However, in spite of the improvement in the illumination the endoscope as a surgical tool clearly did not transcend the experience of the brief succession of enthusiasts involved in its application to neurological surgery.

The Hopkins Advances

In 1960 Harold Hopkins of the University of Reading, England, devised a new lens system with novel features. These overcame the serious optical defects which he had discovered in instruments designed according to the old Nitze system which consisted of a train of biconvex glass lenses spaced out in an air-containing metal tube (Fig. 1). Hopkin's solution consisted of

Fig. 1. The configuration of the lens train of the Nitze system, with bi-convex glass lenses held by spacing tubes which reduce the effective optic diameter

the subsitution of a train of glass rods (the solid rod lens system) now forming much the greater volume of the metal tube, so that there was in fact a series of air lenses in glass (Fig. 2). Light transmission in an endoscopic system is proportional to the square of the refractive index of the transmitting medium. Refractive index is defined as the ratio of the velocity of light transmission in air to that in a different medium. For the glass used in rigid endoscopes this figure is 1.5 the square of which is 2.25. Simply substituting glass for air therefore results in more than doubling the light transmission for a given diameter. The solid rod lens system was able to utilise more of the effective internal diameter of the tube since it did not need stopping down to the centre of the lens system which the Nitze layout required because of spherical aberration. This modest gain in the clear aperture in the solid rod lens system was approximately 1.4 in the radius. However light transmission in such a system is proportional to the fourth power of the radius (1.4 to the 4th power = 4). This figure of 4 when multiplied by the 2.25 gain from the use of glass rather than air results in a total increase of 9 times in the solid rod lens light transmission capacity compared with the Nitze system. In addition the coating of lenses and a

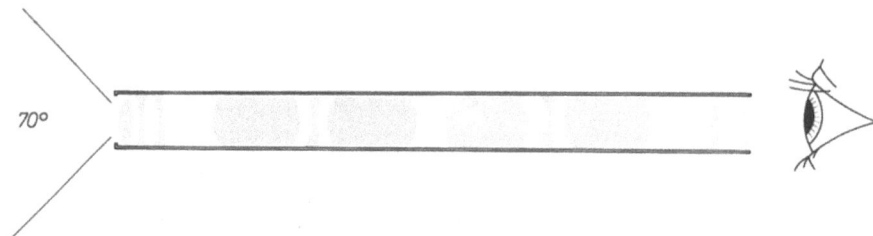

Fig. 2. The Hopkins solid rod lens system which makes greater use of the internal diameter of the containing tube and which avoids problems of the alignment of lenses and the use of spacer tubes

computerised lens layout resulted in a very much more efficient system. The practical consequence is that it is now possible to construct a telescope of high optical quality which is 2.6 mm in external diameter. This instrument is correctly colour balanced, is without major field distortion, and has a much wider angle of view (70 degrees rather than 40 degrees) than other lens systems (Berci and Kont 1969). Since chromatic abberation and ghost images had been done away with the efficiency of light transfer through the tube was so much greater that it was now possible to make this instrument the basis of an endoscopic operating system suitable for use inside the human brain. This telescope forms the basis of a neurosurgical operating system which has made possible the foundation of a new and exciting branch of neurosurgery, namely endoneurosurgery.

The phenomenon of total internal reflection needs some consideration for it is with incongruent fibre bundles that the important business of illumination of the object to be viewed has been achieved. Fig. 3 shows what occurs if the incident ray from material of a higher refractive index (*e.g.* water or glass) passes into a medium of low refractive index (*e.g.* air). If the incident ray encounters the optical interface at more than a critical angle the light is reflected internally. Very little passes through the surface. Successive reflections will continue to take place so that all the light is transmitted inside the fibre. However surface irregularities and contamination, *e.g.* grease (which has a higher refractive index than glass) will seriously impair the capacity of the internal surface to reflect light. Consequently each fibre of high refractive index glass (1.69) is coated with a layer, 2 µm in thickness, of glass of a lower refractive index. However this coating reduces the effective area of the fibre which is responsible for light transmission as none can be transmitted by the cladding. Since the cladding is on the outside of the fibre, it takes up a substantial area so that only 60–70% of the light striking the end of the fibre bundle can be transmitted. A further limitation

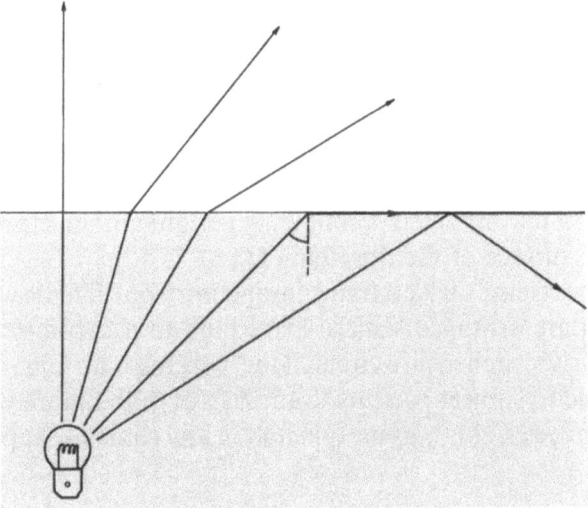

Fig. 3. The principles of refraction and total internal reflection

is that of the packing fraction which is the inherent loss of area in between fibres of approximately spherical configuration when they are packed in a fibre bundle. These limitations can be overcome by increasing the light impinging on the end of the fibre bundle.

When light is transmitted through any medium, losses due to absorption occur. The glass used for fibre lighting transmits ultra-violet and the blue end of the spectrum much less than at the peak of colour transmission which occurs in the yellow. In addition long fibre bundles transmit less light, the losses being proportional to the length.

Light Sources

Illumination has always been a problem. For light to be introduced first into a guide and then into the cavity to be illuminated, it has to be concentrated. It has to be representative of the visible spectrum so that the full range of colours can be presented to the human eye, increasing diagnostic accuracy. Thirdly, as far as possible it has to be cool light.

Light Concentration

In the first Nitze type endoscope the miniaturised incandescent bulb was placed at the "business" end of the endoscope. Light is released from an incandescent filament in all directions and is of course, subject to the inverse square law. This arrangement is wasteful in that many areas are illuminated which cannot be seen. It is clearly more productive to illuminate only the visible area. It is possible to place the incandescent bulb behind a lens which goes some way towards bringing about this field match between illumination and vision. However, when an external light source is used concentration can aim the light beam onto the guide (quartz or fibre-optic) which transmits it further. A parabolic mirror placed around the point source of light, either a filament or an arc, is usually employed. Consequently most light sources employ a concave mirror behind the bulb directing the beam towards a lens which converges onto the fibre cable. The area of this is modest, only a few millimeters square. A certain amount is reflected from the reflective surface of the fibre bundle.

Incandescent lamps have a finite life and burn out. This is why most fibre light sources are arranged with a spare bulb in a capstan which can be instantaneously switched into circuit if the working bulb fails. Since it would be inadvisable to subject patients to the risk of bulb failure when working inside the ventricles, this is a sine qua non of any clinically appropriate light source.

Colour Balance

The gold standard for colour balance appropriate to the human eye is the light of the sun. Many varieties of light source give energy outputs which approximate to daylight but all of them deviate from it in some important way. For instance, fluorescent light is over-balanced at the blue end of the spectrum and an ordinary incandescent filament at the red and yellow end of the spectrum. The xenon arc, a very powerful source, likewise has blue accentuation. Monochromatic light, green, blue, red or yellow, is purest but such coloured light would give a very distorted idea of the normal appearances inside the ventricle.

The *transmitting media* also distort the colour balance of light. For instance ordinary glass produces an accentuation of the bluish end of the spectrum while the glass employed in fibre-optics makes the view appear more red an yellow than is actually present. Light transmitted through water is distorted by preferential absorption of the red and yellow end of the spectrum. The green and blue bands are much better preserved, hence the colour of the sea!

Animal tissue too, has a selective light absorbing power. Haemoglobin in tissue gives light transmitted through it a red tinge. When a fluid collection is seen through tissue it appears dark, as even colourless fluid *e.g.*

cerebrospinal fluid, is a light absorber. When approached with the endoscope the ventricle through the brain appears as a dark loom as the ependyma is approached.

Heat Output

The output of heat is an inescapable consequence of light production and absorption. Fluorescent lights are better in this respect but they suffer from the disadvantage that light output is diffused. A complicated arrangement of mirrors would be necessary to concentrate a fluorescent light source usefully. Similarly a quartz halogen bulb produces more light and less heat from the same wattage of electrical power as an incandescent bulb. The xenon arc output provides a large amount of heat by virtue of the fact that a large amount of energy can be pumped into it, although watt for watt it is not especially burdened with an unwanted heat spectrum. The usual device for overcoming this undesirable heat output is a dichroic mirror. The normal arrangement is that the spherical or parabolic mirror around the light source is coated. This reflects the visible spectrum but allows much of the heat generated to travel through the mirror to be dissipated in the housing of the lamp. These dichroic mirrors are moderately efficient but the intensity of light required still carries with it a sizeable quantity of heat. One of the best methods of dissipating this is to provide a heat sink at the end of the fibre optic lighting cable. This is made of metal, not only to fit accurately into the housing where the light can be concentrated upon it, but also to dissipate heat production. However even though these two devices are used a large quantity of heat is inescapably transmitted into the bundle with the possibility of damage to the flexible fibre lighting cable. When the xenon arc is used for more than a few minutes the operator can by palpation be made aware of how much heat production is being transmitted and lost through the cable. Heat production at the endoscope is not so important since the butt or viewing end of the endoscope is constructed largely of metal which again acts as a heat sink, as does the thin stainless steel tube in which the rigid endoscope is enclosed.

Fibre-light sources for clinical use are now fairly reliable. An ordinary incandescent bulb source for general use without photography is to be found in many operating theatres. Their output is perfectly adequate for inspection purposes. For photography however, a xenon arc arrangement, much more expensive is usually employed. This has special arrangements to strike and modulate the arc. For instance, the Storz light box which employs both flash, xenon and incandescent bulbs, has a control for varying intensity of the xenon arc. In practice, all three outputs provided are of similar intensity. The xenon arc can be used only in relatively short bursts of a few minutes or so at a time before over-heating becomes a problem.

Manipulations

It is possible to use the telescope alone, advanced through the brain into the cerebral ventricle, access to the cerebral cortex being given by a small burr hole. The bare telescope is in the same size range as an ordinary brain cannula, but it is slightly more awkward to manipulate since it has attached to it a rather heavy lighting cable. The eye is held to the eyepiece, and as the instrument is advanced through the brain, a uniform field of light with flashes of red can be seen, then a darkening as the objective lens nears the ventricular wall, and then a sudden clear view of the inside of the ventricle and its contents appears as the end of the telescope bursts through the ependyma. Used in this mode, fragments of brain tissue occasionally remain adherent to the end of the objective lens, seriously impairing the field of view. For operating inside the ventricle an operating sheath and trocar (at present 4.5 mm in diameter) is introduced down the track which the telescope has made through the brain. The trocar is withdrawn and the operating instruments are then insinuated into the sheath alongside the telescope. An electrode, a rigid aspirating cannula, an accessory light guide of Crofon (perspex sheathed in methacrilate) and a plastic cannula for perfusion of physiological saline into the ventricles can all be accommodated in the 4.5 mm sheath alongside the telescope. The rigid cannula can be replaced by an endoscopic scissors of the same dimensions, and when all the accessories are removed, a grasping forceps for retrieval of biopsy material and of shunt cannulae which have become detached into the ventricle. However, due to the lack of stereoscopic vision through the endoscope, and coupled with the effect of the inverse square law, it is surprisingly difficult to carry out grasping manipulations since the judgement of distance is very much poorer than would be apparent at first sight.

For babies and infants the sheath can be introduced directly into the ventricle by scalp puncture through the fontanelle or through the lambdoid suture (Griffith 1975). This is done by fitting it with a sharp trocar and by making the puncture through a small incision in the scalp, approximately 5 mm in length. The trocar and cannula are advanced through the suture, dura and brain, and when the ventricle is penetrated the trocar is withdrawn and the telescope and operating equipment are introduced. When inspection alone is needed a smaller sheath (3.7 mm in diameter externally) can be used. The ventricular pressure is maintained through a drip of physiological saline or artificial cerebrospinal fluid taken through a temperature-controlled warming bath.

Neurosurgical operations on the head are usually carried out with the patient on a headrest specially designed to give close access to a large proportion of the head. The elbow and/or wrist support which is necessary for fine work either with microscope or endoscope, is usually provided by a specially designed chair. With the endoscope however, this provides a somewhat restricted posture for the surgeon. Consequently we have found, that especially for infants and children, it is preferable to place the patient on an ordinary operating table. The head is supported in the semi-prone position by an evacuated bean bag or moulded sponge in the case of infants. This presents the occipital region to the operator. The operator's elbows can then be supported on the operating table. This means that the patient's body is supported on the opposite side of the operating table from that at which the operator sits. When the telescope is grasped in one hand by the fibre-light source attachment then the other manipulations necessary such as of the diathermy electrode, can be carried out by the other hand. The support of both elbows ensures safety.

When the fontanelle is entered the patient is then in the supine position. The head is placed at approximately 12–15" from the end of an ordinary operating table. The flat surface is then available for elbow support. Naturally the head is stabilised with a ring or evacuated bean bag.

At present the endoscopic system is used almost exclusively inside the cerebral ventricles, or cysts or diverticula from these cavities. It is possible (I have performed this in the cadaver) to inspect the subarachnoid space of the cisterna magna and at the cranio-cervical junction, and also to take photographs of this region and also of the spinal nerve roots when the trocar with its ensheathing cannula is used as a lumbar-puncture needle.

Endoneurosurgery and Hydrocephalus

Briefly, cerebro-spinal fluid is produced by the choroid plexus, although it is not yet clear as to whether the choroid plexus situated inside the ventricles of the brain enjoys a monopoly position for the seat of this process by active transport. It is still unclear as to whether there is much variation in the rate of secretion of cerebro-spinal fluid. The few measurements which have been made in human beings both in health and in hydrocephalus (Lorenzo, Page and Watters 1970) have been carried out by a perfusion method which entails a period of several hours for equilibration and which would therefore be unsuitable for the discovery of short-term variations. Again, although drugs such as acetazolamide and Digoxin can depress the formation of cerebro-spinal fluid, it is in the absorption mechanism for cerebro-spinal fluid that the defects leading to hydrocephalus can be found (Fig. 4). Hydrocephalus is seen largely as a defect of absorption so that a relatively stable secretion rate matches an absorption rate at an increased pressure. Since the alteration of absorption rates could not be directly improved except possibly by valve drainage (Milhorat *et al.* 1971) the therapeutic manoeuvre which holds out most hope of controlling the patho-physiology of hydrocephalus lies in cutting down the secretion rate to match the absorption mechanism at a greatly decreased pressure.

Choroid plexus coagulation using unipolar diathermy has been carried out in hydrocephalic infants over the last fourteen years. An illustrated result is shown in Fig. 5. In this patient the acute phase of hydrocephalus followed meningitis controlled by antibiotics. The advance in head size was rapid, but there was clear/control after choroid plexus coagulation, and at the age of 2 years the occipito-frontal circumference of the skull appears to be

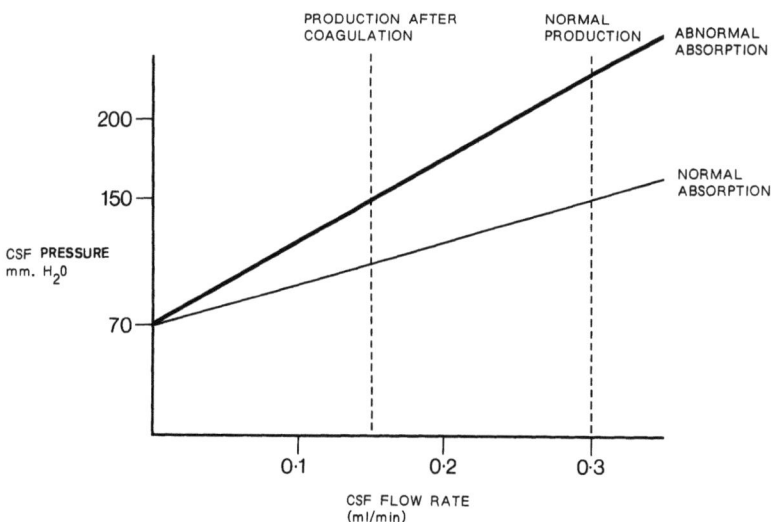

Fig. 4. Diagram to illustrate what are believed to be the main facts of the pathophysiology of hydrocephalus. Secretion of cerebrospinal fluid is largely independent of intracranial pressure, but absorption is pressure-dependent. Hydrocephalus results when the absorption process is less pressure-sensitive, and a greater cerebrospinal fluid pressure is required to achieve absorption of the amount secreted. Choroid plexus coagulation achieves a normal pressure on the steeper hydrocephalic absorption curve at approximately 50% of normal secretion rate

following the 50th percentile, in marked contrast to that at the stage of active hydrocephalus. By the usual clinical criteria this child's hydrocephalus has been completely controlled. We are at present treating all children with communicating hydrocephalus by this manoeuvre, coupled if necessary by the insertion of a ventriculo-atrial shunt which can be switched on and off. The purpose of this valve, which we always hope will be temporary only, is to improve the external cerebro-spinal fluid circulation by increasing the capacity of the absorption mechanism. Children who suffer from genuine aqueduct stricture have not been considered controllable by choroid plexus coagulation but with water soluble positive contrast ventriculography only a small minority of children with hydrocephalus appear to have a genuine aqueduct stricture (but see later).

The selection of patients for this manoeuvre has been according to the usual criteria, namely an infant with a cranial circumference progressively diverging from the curve of normal development with grossly enlarged ventricles (judged in the earlier patients by fractional ventriculography and in the vast majority in this series by CT scan) with an absence of superficial

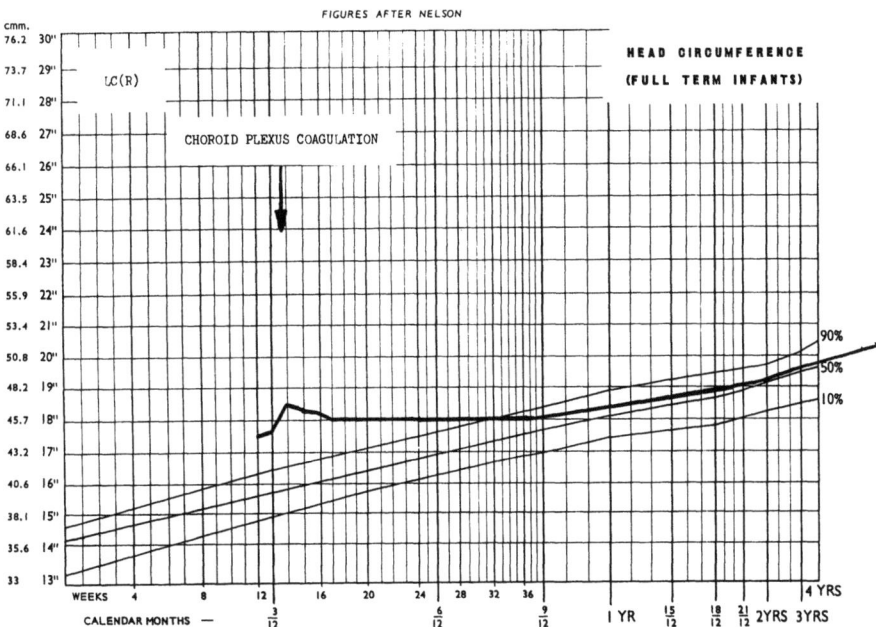

Fig. 5. Normal head growth curve defined by parallel continuous lines on a logarithmic time scale. Patient's head expansion curve defined by dashed line. Arrows show time of choroid plexus coagulation, which resulted in arrest of head expansion and subsequent rejoining of normal growth curve on 50th percentile

cerebro-spinal fluid spaces over the cerebral hemisphere, in the sylvian fissures and between the hemispheres as seen in the CT scan. Intracranial pressure monitoring for periods of longer than 24 hours has been occasionally used in some of the older patients in this series but has not been particularly informative. Behaviour change with deterioration of performance has been noted in many patients. Where such functional evidence has been lacking we have been reluctant to embark on either choroid plexus or valve surgery without a further period of observation. Over the same period a small number of patients have been treated by valve surgery alone, the choice being determined by the exigencies of a hardpressed clinical service and absences on holiday etc., rather than any discernible difference in the disease pattern. Some patients have been referred by colleagues after previously unsatisfactory and complicated valve treatment of hydrocephalus. The vast majority of procedures (95%) have been carried out personally.

The series comprises 71 patients operated upon in the years 1972–1983 inclusive. One patient died on the second post operative day of haemorrhage. The remaining 70 have been followed up and fall into five categories. These are:

1. Patients whose hydrocephalus has been controlled by choroid plexus surgery alone and have never needed a shunt.
2. Patients who have had a shunt at some stage, sometimes pre-operatively, but in whom the shunt has been removed.
3. A shunt which has repeatedly been demonstrated by all the usual clinical criteria to be non-functioning.
4. A doubtfully functioning shunt.
5. Definitely shunt dependent.

The patients have been divided into three broad categories of patients with myelomeningocoele, those with communicating hydrocephalus, and those with obstructive hydrocephalus, the vast majority due to aqueduct stricture. The results are set out in Table 1.

Table 1

Category	Meningomyelocoele	Communicating Hydrocephalus	Obstructive Hydrocephalus	Total
1	5	11	3	19
2	1	1	1	3
3	6	6	1	13
4	2	2	2	6
5	12	15	2	29
Total	26	35	9	70

The results show that 30% of patients (categories 1 and 2) can be controlled with choroid plexus coagulation alone. If category 3 patients are included, that is to say, those in whom a shunt is probably not functioning, the success rate rises to 49%. These results have given us confidence to persist with the technique. Although the numbers are small the results in aqueduct stricture do not differ from the larger groups.

After choroid plexus coagulation blood and protein in sizeable amounts are found in the cerebro-spinal fluid. Both these constituents are known to promote communicating hydrocephalus by interference with function in the arachnoid granulations with impairment of cerebro-spinal fluid absorption. A period of open drainage for several days after the procedure, or the subsequent insertion of a shunt, has not much improved the results. However perfusion with artificial cerebro-spinal fluid for 48 h after choroid plexus coagulation heralds some hope of avoiding such damage to CSF absorption capacity, with an improvement in results. The numbers are so far too small and too recent to enable a conclusion to be reached about the success of this adjunctive treatment.

In patients with obstructive hydrocephalus due to aqueduct stricture it is sometimes possible to unblock the stenotic aqueduct by the Seldinger technique. The burr hole must be placed well anterior to the coronal suture in order to line up the foramen of Monro with the entrance to the aqueduct. The Seldinger wire is introduced down the aqueduct and a check x-ray taken to ascertain correct positioning. The tube for intubation is then manoeuvred over the wire and propelled along it through the strictured portion by nudges and pushes with the endoscope itself. When the position is satisfactory the Seldinger wire is then withdrawn and a positive contrast ventriculogram carried out to ascertain entry of dye into the fourth ventricle. Only small number of cases have been attempted by this method, however, and inherent dangers especially to the control of eye movements render it of less than general utility. More effectively, a similar technique can be used through the thinned out floor of the third ventricle—a posterior third ventriculostomy. Although an endoscope may be used for this technique it has been satisfactorily performed without endoscopic aid (Loew *et al.* 1981, Jaksche and Loew 1986). Indeed, the results quoted in this last paper of 80% cure of non-tumoural aqueduct stenosis by posterior third ventriculostomy seem somewhat superior to the results of choroid plexus coagulation.

Intraventricular Tumour Biopsy

The other sizeable use for the endoscopic technique at present is for biopsy of tumours impinging upon the ventricular system. These are often tumours which metastasise via the cerebrospinal fluid pathways such as pineal tumours or ependymomas of the cerebral ventricles. CT scan will often guide the operator so that he can decide whether to introduce the endoscope into the ventricle from behind or more frequently to visualise the anterior third ventricle via the foramen of Monro. With an anteriorly placed burr hole so that the straight endoscope can be introduced via the burr hole near the midline into the dilated lateral ventricle through the foramen of Munro and across the dilated third ventricle, it has been possible to biopsy even a tumour in the fourth ventricle using this technique. Biopsy forceps encircle the endoscope and all fit inside the 4.6 mm ensheathing cannula introduced with a trocar. When the biopsy forceps are opened the cups comes to lie just above the centre of the field of view. In this way the tissue can be visualised, approached, and a bite taken. Bleeding then usually ensues but can often be controlled by diathermy. A stream of warmed artificial cerebrospinal fluid is directed alongside the unipolar diathermy electrode. However if the bleeding is too great for this to be successful, transventricular irrigation will often allow, after a further period, another bite for biopsy or for haemostasis.

Inspection of the third ventricle will, even without a biopsy, often show subependymal or ependymal deposits from a pineal tumour usually in the posterior part of the third ventricle. Gliomas of the basal ganglia in young children, often benign astrocytomas with cysts, can sometimes be marsupialised into the lateral ventricle as well as biopsied. It is our practice with a posterior third ventricular tumour to carry out an inspection of the posterior third ventricle before carrying out any therapeutic manoeuvres such as a shunt. Information about cerebrospinal fluid pathway metastasis can thereby be gained even when unsuspected by CT scan. A decision as to whether to shunt externally into the venous or peritoneal system can then be modified to perhaps providing an internal shunt (ventriculo-cisternotomy).

Of course other intraventricular processes can be biopsied. We have recently carried out a successful biopsy on a patient with cerebral sarcoidosis. The ependyma was studded with raised ependymal patches sometimes congregating around septal and sub-ependymal veins. The Foramen of Munro appeared to be blocked by this process. With the grasping forceps, biopsies were taken, histology showing the characteristic granulomatous picture of sarcoidosis. As the diagnosis in these patients is often difficult and as the ventricles are dilated, endoscopic biopsy seems to be a step forward in the rigorous tissue diagnosis of this disease, so often doubtful or lacking.

Endoscopic Photography

Although it is possible to take still photographs and to take cine and video films through the endoscope, the technical demands are such as to make it difficult to use these forms of visual recording routinely. The important factors controlling the ability to photograph inside the ventricles are set out in a recent paper by Tutty and Russell (1986). These are those of illumination, magnification, and film characteristics, especially grain.

It is possible to increase the light available for photography by pumping light from a xenon arc through the fibre lighting bundle of the endoscope. Secondly accessory light fibres are sometimes useful. By means of a teaching attachment 90% of the image light can be diverted to the camera, the human eye being somewhat more sensitive, requiring less light. A second important factor in illumination is the distance of the endoscope from the illuminated object, since the inverse square law applies. All the light which goes to form the image must fall on the small collecting lens, less than 2 mm in diameter.

If this field or collecting lens is pushed closer to the object to be photographed, it will automatically be illuminated more densely although of course at the same time the image will become magnified.

The magnification of the image is such that an image size on film of approximately 10 mm diameter has been acceptable at a shutter speed of 1/30 second or less. The need for this fast exposure is to preserve image clarity in the face of pulsation and movement intracranially. Although such an image can be enlarged further, doing so loses detail.

The grain of the film is related to the film speed. Very fast film (around 1000 ASA) is used. A faster film is more grainy and therefore gives a less precise image. In addition the optical axis of the camera has to be accurately aligned over that of the teaching attachment and endoscope, with automatic exposure control necessary because of the enormous range of light intensities. Fortunately most automatic exposure control cameras take their meter field from the centre which is the part illuminated by the endoscope. Problems with sterility can usually be overcome. Three photographs of ventricular anatomy and pathology (Figs. 6, 7, and 8) show the possibilities of the technique.

Other Diseases

It is possible to operate on colloid cysts of the foramen of Monro via the endoscope. The cyst can be incised, the contents (if liquid) can be removed by suction and the wall can be shrunk and coagulated with the unipolar diathermy (Powell *et al.* 1983). However this technique has not yet reached the level of microscopic open surgery where the cyst can be completely and safely removed without risk of recurrence. However with light guides and laser surgery it seems possible to remedy this failing in the near future so that both the cyst and contents, even if solid, can be vapourised endoscopically and a permanent cure obtained by endoscopic means alone via a burr hole.

Before the advent of the CT scan, Ehrenberger (1981) and Oppel (1981) inspected the cerebellopontine angle with the endoscope via a laterally placed craniectomy in order to be certain as to the presence of an acoustic tumour so that full exploration of the angle could then proceed. However the advent of more sophisticated CT scanning, coupled with air contrast, has provided this information non-invasively so that the need for this technique has receeded. However the same technique can be used for the assessment of vascular loops in the cerebellopontine angle both in trigeminal neuralgia and hemifacial spasm.

The use of the fore-oblique endoscope in transphenoidal surgery to ascertain whether more tumour is present beyond the field of the operating microscope has been used but not to much effect. Blood tends to get on the lens and for oblique viewing is much more difficult to interpret than straightforward viewing.

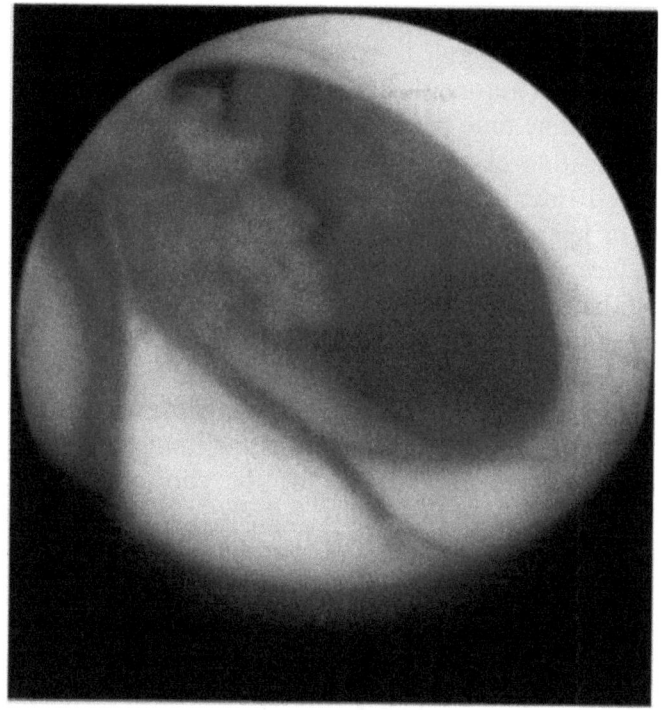

Fig. 6. Foramen of Monro looking into third ventricle with choroid plexus septal and internal cerebral veins and connexus interthalamicus

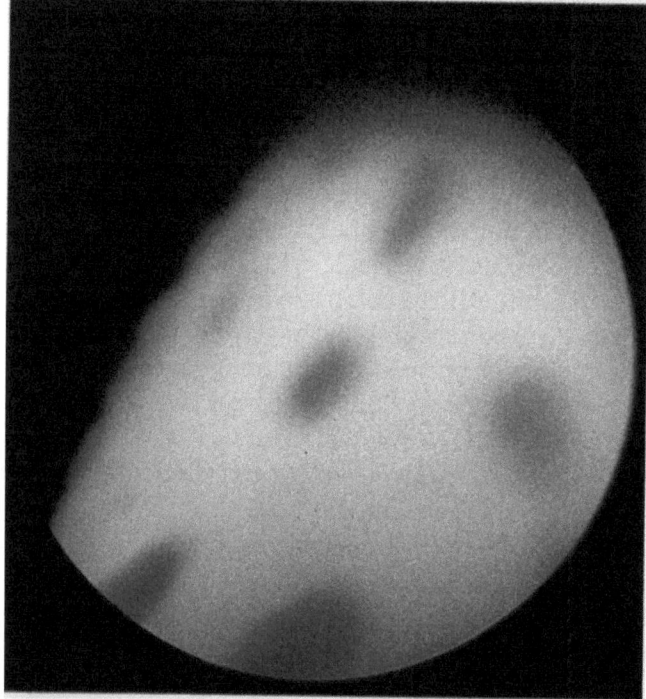

Fig. 7. Brown granular ependymitis. These slightly raised pigmented excrescences are found on the ventricular wall after intraventricular haemorrhage

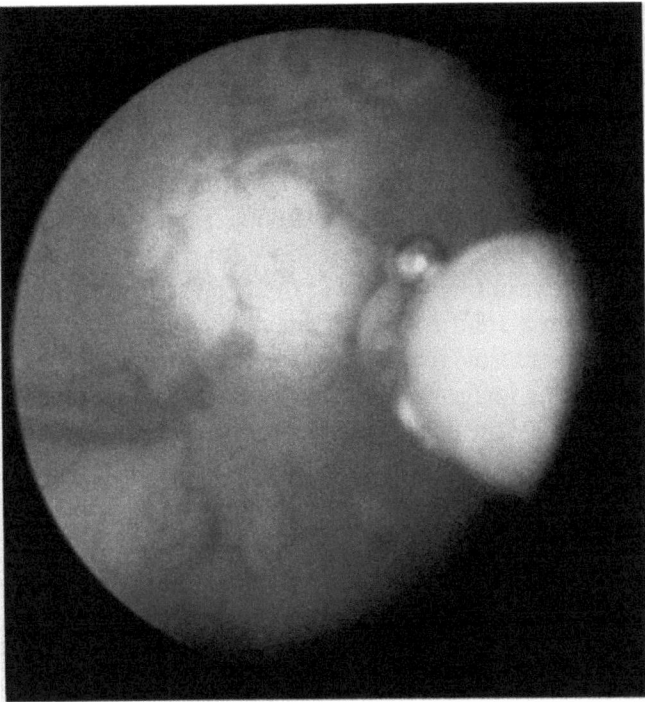

Fig. 8. Fulgaration with the unipolar electrode. A bubble is seen formed on the electrode. The bleeding point has blanched white

Small Flexible Endoscopes

The congruent fibre bundle, which Hopkins again pioneered, has had a large impact in other fields of medicine. Gastro-enterology for example, has been revolutionised by the ability of a flexible endoscope to penetrate the tortuosities of the gastro-intestinal tract; they give good service in this area. Why then is it not possible with miniaturisation to carry out a similar function for the tortuosities of the cerebrospinal fluid pathways and ventricular system? The answer lies in the way the image is formed in a congruent fibre bundle and in the nature of light itself. The image obtained by flexible fibre transmission endoscopes is made of a series of discrete elements each corresponding to an optically sheathed fibre through which the light is transmitted by total internal reflection. The light intensities do not vary over the surface of each element so that the only difference in the light emerging from one element and its neighbour is the intensity. This piecemeal picture can be overcome by increasing the number of fibres. Many thousands are needed to form an endoscope with acceptable resolving power. For instance a bundle containing 10 000 fibres has only 100 in each row. When miniaturisation is necessary, as in the nervous system, then each element could be made smaller. However there rapidly comes a point at just less than 20 µwhere the wave-length of the light governs its ability to enter a glass fibre element. Clearly if light cannot enter the fibre it cannot be transmitted through it to form an image at the other end. Thus to occupy a space comparable in size to the Hopkins solid rod ventriculoscope of a diameter 2.6 mm, many less than 10 000 elements can be incorporated when packing difficulties and lubrication for flexibility are taken into consideration. The resulting image will then be vastly inferior in quality to that obtained from a solid rod lens system of comparable size.

When a gastro-enterological endoscope is examined it will be seen that the light transmitting image transmitting bundle forms a very small fraction of its total diameter. More important, the control wires needed for flexibility take up a great deal of space. When the thick plastic sheath is added it will be seen that the space requirement amounts rapidly.

There are three further matters in which the flexible endoscope performs less satisfactorily than the solid rod lens system. One is the acceptance angle for the field lens. This is a good deal narrower in the case of the flexible endoscope in order that the light can enter the fibres. Secondly the colour balance of the system is very much poorer with the flexible endoscope, something which considerably decreases the ability of the surgeon to recognise colour differentials between tissue. Thirdly the flexible endoscope is very much more difficult to sterilise. In gastro-enterology this is not of much consequence but in order to meet acceptable standards for use in the nervous system gas sterilisation (which may decrease the working life of the instrument) is required. This decreases rapid and easy availability of the instrument which is of great value in neurosurgery. Lastly the life of the flexible instrument is limited since fibre elements break and stand out as spots in the image. The durability of a solid instrument is vastly superior in this respect.

However the inherent advantages of flexibility are very important. The dividends which could be gathered were a miniaturised system with an acceptable image quality to become available if the diameter problem could be circumvented (possibly by external magnetic control rather than by wires around the bundle) a new impetus would be given to "flexible endoneurosurgery".

Sterilising Techniques

Surgical instruments are of no use unless they can be adequately sterilised. Because of the complicated construction of endoscopes, with air, glass, lens cement, and metal all in close conjunction, this need has in the past given rise to difficulties. For urological use, where a succession of patients are cystoscoped, various manoeuvres such as the use of low-temperature steam have been used. When a surgeon consults a bacteriologist with the question, "What is the best method of sterilising an endoscope?", they wary bacteriologist will reply, "Just how sterile do you wish your endoscope to be?" In the new field of endoneurosurgery it is not possible to be certain, except that the effects of infection, were it to occur after endoscopy of the brain, are probably more devastating and far-reaching than in any other body system. However, in order to make best use of a diagnostic and therapeutic tool, such as an endoscope, it must be rapidly available so that it can be used in unexpected circumstances as well as in those cases where its use can be foreseen. Consequently we have tried to use and develop methods which do not damage the endoscope, but which give a very high quality of sterilisation in as short a time as possible. We have used three main methods.

(1) Low-temperature steam sterilisation, with and without formalin. For our first 30 or so ventriculoscopies the endoscope was sterilised in low-temperature steam and formaldehyde, the cycle involving a temperature of 70 degrees C for 15 min (Alder, Brown and Gillespie 1966). At the end of this time it was noticeable that the light transmission characteristics of the instrument had altered so that blue light was transmitted better than the red end of the spectrum, and we changed to:

(2) Activated glutaraldehyde (Cidex-Arbrook Products). In order to kill spores this method requires immersion of the endoscope for $2\frac{1}{2}$ h and then a wash and rinse in sterile saline before use. Although easier to organise than low-temperature steam in our Unit, this still made demands on the theatre staff when a baby was to be operated upon, as they should be, at the start of an operating day. Consequently we have gone over to:

(3) Methanol-hypochlorite. This system was developed by Dr. Kelsey of the Public Health Laboratory Service (Kelsey, Mackinnon and Maurer 1974) as a result of discussion at a meeting of the British Society for Medical Endoscopy. We have now used the system for 11 years and have found it rapid and effective. The instruments are immersed for a period of 15 min in a mixture of equal parts of methanol and hypochlorite, rinsed in saline, and then are ready for use. This seems a very much more practical method than any other so far devised.

Future Developments

The cisterna magna, where the endoscope is introduced by cisternal puncture either in the mid line or from laterally, is a field for further study. The ability of the surgeon to introduce coagulating electrodes for pain control in trigeminal neuralgia (spinal tract and nucleus) and into the spinothalamic tract, holds out a hope of a more precise form of percutaneous pain control surgery than has so far been available. However the cisterns here are often fairly small so that the endoscope is usually close to the structures observed. In order to facilitate this kind of surgery an experimental neuroscope was constructed, much smaller in dimensions than the present 2.6 mm diameter instrument. However technical problems in flexibility made this less useful than would at first be apparent so that further miniaturisation would probably not hold out enormous benefits. A field which is still as yet unexplored is the use of the endoscope as a lumbar puncture instrument so that the roots of the cauda equina can be stimulated directly, and if necessary coagulated in pain control in a much more precise way than is possible for the instillation of alcohol or phenol. As both of these agents frequently result in sphincter disturbance, again such a technique might well be helpful when combined with feedback from stimulation evoked paraesthesiae and possible temporary lesion making.

Artificial Cerebrospinal Fluid Solution

When ventricular volumes are high and the amount of perfusate required to top up the ventricle or to clear wisps of blood is small, then ordinary physiological saline solution has been found to be suitable. Little systemic reaction is subsequently seen in these patients. However if a large volume of perfusate is used, for instance in the replacement of cerebrospinal fluid volume if more than minimal bleeding takes place, then the ordinary saline solutions are capable of inducing toxic reactions. This is because their osmotic pressure, membrane active ion concentration, and pH and CO^2 are not within the physiological range for cerebrospinal fluid. Consequently we have developed a simple two part solution which gets over many of these difficulties. Basically it is cerebrospinal fluid without glucose or protein. Its pH is around 7.4, its osmotic concentration is 290 mmol/l and its bicarbonate concentration and CO^2 content is physiological. It contains the requisite concentrations of sodium, potassium, calcium, magnesium and chloride. It is prepared in two parts, a calcium solution being added to the basic solution which is then ready for use.

When the need arises, such as when substantial bleeding occurred on one occasion, rapid infusion and drainage into and out of the ventricle to the extent of half a litre in ten minutes, is safely carried out. We are using the solution for irrigating blood and protein from the ventricles after choroid plexus coagulation as previously described. Naturally this solution has a variety of uses in neurosurgery and other uses for it readily come to mind. These include the inflation of the brain by the lumbar route after the removal of extensive liquid subdural haematoma and irrigating the blood from patients with ruptured intracranial aneurysms or neonatal basal ganglia haemorrhages. Commercial firms have in the past thought that making solutions of this kind available has been economically unviable but the existence of a steady demand for this in neurosurgical departments will no doubt induce the economics to alter.

References

Alder VG, Brown AH, Gillespie WA (1966) Disinfection of heat sensitive material by low temperature steam und formaldehyde. J Clin Path 19: 83–89

Berci G, Kont LA (1969) A new optic system in endoscopy with special reference to cystoscopy. Br J Urol 41: 564–571

Ehrenberger K (1981) Reversible functional damage of VIIIth cranial nerve in arachnopathia pontocerebellaris. In: Samii M, Janetta P (eds) The cranial nerved. Springer, Berlin Heidelberg New York, p 575–578

Grant FC, Fay T (1923) Ventriculoscopy and intraventricular photography in internal hydrocephalus. Report of case. J Am Med Ass 80: 461–463

Griffith HB (1975) Technique of fontanelle and persutural ventriculoscopy and endoscopic ventricular surgery in infants. Child's Brain 1: 359–363

Guiot G, Rougerie J, Fourestier M, Fournier A, Comoy C, Vulmiere J, Groux R (1963) Une nouvelle technique endoscopique. Explorations endoscopiques intracrâniennes. Presse Med 71: 1225–1228

Hopkins HH Br Pat No 954629: US-Pat No 3247906

Jaksche H, Loew F (1986) Burr holes third ventricle cisternostomy. An unpopular but effective procedure for treatment of certain forms of occlusive hydrocephalus. Acta Neurochir (Wien) 79: 48–51

Kelsey JC, Mackinnon IH, Maurer Isobel M (1974) Sporicidal activity of hospital disinfectants. J Clin Pathol 27: 632–638

L'Espinasse VL (1943) In: Davis Neurological surgery, 2nd edn. Lea & Febiger, Philadelphia, p 442

Loew F, Jaksche H, Neuenfeldt D (1981) Ventricle cisternostomy by puncture perforation of the floor of the third ventricle. Indications, technique and late results in cases of occlusive hydrocephalus. Acta Neurochir (Wien) 57: 138

Lorenzo AV, Page LK, Watters GV (1970) Relationship between cerebrospinal fluid formation, absorption, and pressure in human hydrocephalus. Brain 93: 679–692

Nitze M (1887) Veränderungen an meinen elektroendoskopischen Instrumenten zur Untersuchung der männlichen Harnblase. III. Monatsschr ärztl Polytechn 3: 60

Ogata M, Ishikawa T, Horide R, Wamatabe M, Matsumura H (1965) Encephaloscope—basic study. J Neurosurg 22: 288–291

Oppel F, Mulch G, Brock M, Zuhlke D (1984) Indications and operative technique for endoscopy of the cerebellopontine angle. In: Samii M, Jannetta P (eds) The cranial nerves. Springer, Berlin Heidelberg New York, pp 429–437

Powell M, et al (1983) Isodense colloid cysts of the third ventricle: a diagnostic and therapeutic problem resolved by ventriculoscopy. Neurosurg Sept 13 (3): 234–237

Putnam TJ (1934) Treatment of hydrocephalus by endoscopic coagulation of the choroid plexus. Description of a new instrument and preliminary report of the results. N Engl J Med 210: 1373–1376

Scarff JE (1936) Endoscopic treatment of hydrocephalus. Description of a ventriculoscope and preliminary report of cases. Arch Neurol Psychiat 35: 853–860

Scarff JE (1970) The treatment or non-obstructive (communicating) hydrocephalus by endoscopic cauterisation of the choroid plexuses. J Neurosurg 33: 1–18

Torrens MJ, Griffith HB (1974) The control of the uninhibited bladder by selective sacral neurectomy. Br J Urol 46: 639–644

Tutty S, Russell T (1986) J Audio-visual media in medicine (in press)

Evoked Potential Monitoring in Neurosurgical Practice

L. SYMON, F. MOMMA, K. SCHWERDTFEGER, P. BENTIVOGLIO, I. E. COSTA E SILVA, and A. WANG

Gough-Cooper Department of Neurological Surgery, Institute of Neurology, London (U.K.) and Department of Neurosurgery, Saarland University Medical School, Homburg/Saar (Federal Republic of Germany)

With 15 Figures

Contents

1. The Use of Somatosensory Evoked Responses	26
Somatosensory Evoked Responses in Vascular Diseases of the Brain	26
Methods	28
Electrodes and Montage	28
Stimulation	29
Recording Apparatus	29
The Measurement of SSEP a Prognostic Guide in Subarachnoid Hemorrhage	30
The Relationship Between Conduction Time, Prolongation and Clinical Deterioration	34
The Use of Somatosensory Evoked Responses as a Monitor During Intracranial Aneurysm Surgery	37
Techniques	37
SSEP Recording as a Monitor During Temporary Vascular Occlusion	40
Somatosensory Evoked Potential Recording and Use of Somatosensory Evoked Responses in Head Injury Management	43
Clinical Factors Related to Outcome	45
SEP Related to Outcome	47
Progressive Change in SEPs	49
2. Visual Evoked Response Monitoring	51
3. Methodology	52
Recording Technique and Wave Identification	54
VEP as a Guide to Manipulation of the Optic Pathways During Surgery	55
Conclusions	57
4. The Use of Brain Stem and Other Auditory Evoked Potentials	58

Brain Stem Auditory Evoked Potentials in Head Injury	58
Use of Brain Stem Evoked Responses During Posterior Fossa Surgery	59
Electrocochleographic Monitoring During Acoustic Neuroma Surgery	61
The Technique of Peroperative Electrocochleography	61
5. Spinal Cord Monitoring	64
References	65

The functions of the central nervous system are accurately reflected in the generation of electrical impulses. Modern electronic techniques have enabled us to record and analyse a vast number of electrical signals from the central and peripheral nervous system and to attempt a correlation of the form and consistency of these signals with a variety of disease states. For many years the principal effort in understanding the brain's electrical signals lay in an attempt to unscramble the complexity of the electroencephalogram, and indeed immensely valuable advances in the understanding of the function of the nervous system was made by such efforts. From the neurosurgeon's point of view, however, the greatest advance in electronic technology was the application of event related potentials, potentials which could be produced by an externally applied stimulus to a normal pathway, and the behavior of the central and peripheral nervous systems to such stimulation recurrently assessed. It is now possible to challenge the function of all the major afferent systems, somatosensory, visual, and auditory and to monitor their activity at various levels throughout the CNS. Increasing interest has now been shown in the evocation of motor activity by direct stimulation of the CNS and over the next few years we may expect increasing sophistication, both in the techniques for stimulation and of analysis which will render their application to neurosurgical circumstances both easier and more widespread.

This review addresses the principle current techniques in use in neurosurgical units in Europe. As such it is necessarily biased by the interest of its authors and the interested reader will be aware that some challenging aspects of the subject developing in clinics throughout the world have scarcely been mentioned. Nevertheless, the material presented here indicates the proven utility of evoked responses in clinical neurosurgery and it is to be hoped that more widespread communication of individual clinic's experience will lead to dissemination of information in this rapidly expanding field.

1. The Use of Somatosensory Evoked Responses

Somatosensory Evoked Responses in Vascular Diseases of the Brain

The electrical activity of the brain has been related to regional cerebral blood flow both in man[69, 79] and in primates[5, 10, 33]. Clearly established thresholds exist for the brain's electrical function in the primate at regional cerebral blood flow around 15 ml/100 g/min. Below this level electrical activity is completely suppressed. As cerebral blood flow continues to fall there is a further lower threshold of flow in the region of 10 ml/100 g/min at which ionic homeostasis is disrupted, the first detectable movement being a flux of potassium ions from the intracellular to the extracellular space, succeeded promptly thereafter, at a critical level of extracellular potassium, by a movement of calcium from the extracellular to the intracellular space[5, 8, 9, 32, 70]. It was soon clear that persistence of blood flow levels below the level of disruption of ionic homeostasis would inevitably result in infarction[59, 73] and more recently it has been demonstrated that more protracted blood flow levels below the electrical threshold may also result in infarction[7]. In neurosurgical terms, however, the development of acute ischemia is marked by first, a failure of electrical activity, and second, a failure of ionic homeostasis. These two thresholds may be geographically definable, for example with middle cerebral occlusion in man, as in the primate, a core of ischemic tissue whose size and extent depends on the availability of collateral circulation, will rapidly fall below the ionic threshold and unless reperfusion is prompt, the tissue will infarct. Where depression of blood flow is less severe, as for example in the peripheral distribution of the middle cerebral artery, electrical suppression may be present but infarction may be deferred for longer periods of time, enabling for example temporary vascular occlusion to be carried out with safety. In the acute phase of cerebral infarction the area between these two thresholds may be geographically definable and has been described as the ischemic penumbra[4].

Electrical activity, therefore, holds out promise as an early warning for the development of serious brain ischemia.

Initial attempts to determine the electrical activity of cortex by the use of focal electroencephalographic methods have proved disappointing. There is no doubt that regional suppression of integrated EEG activity can be demonstrated, but these changes are often slow, complex and difficult to analyse and the technique of evoked electrical activity has nowhere proved of greater use than in the analysis of brain ischaemia. Our own work was stimulated by experiments on primates, clearly demonstrating the rapid failure of evoked response to critical levels of brain ischaemia, and although initial use was made of evoked potential amplitude, the variability of amplitude recordings in clinical circumstances soon made it evident that amplitude recording was of less than adequate reliability. Thus, from day to day, amplitude recording is much dependent upon scalp thickness which may vary in the postoperative case, much dependent upon the level of alertness of the patient, and on the presence or absence of subdural fluid. The key technique, therefore, has followed the work of Hume and Cant[36] in New Zealand in head injuries, where the measurement has been of the time of transmission of electrical impulses through the nervous system. Hume and Cant showed that evoking somatosensory response by median nerve stimulation at the wrist, one could time the arrival of the impulse at the gracile and cuneate nuclei by the appearance of the wave N 13 or N 14, and thereafter its passage through the central nervous system itself to its arrival at the somatosensory cortex, the peak N 20 indicating the arrival of the primary wave. The N 14, N 20 interval became known as central conduction time. It has the great advantage of being independent of peripheral phenomena such as the temperature of the limb, the presence or absence of drips in the arm, arm length, and so forth, and of enabling comparison between one hemisphere and the other. This technique has been applied by a variety of groups [18, 22, 37] and is now well accepted.

Methods

SEP recordings have been recorded from median nerve stimulation at the wrist using surface electrodes.

Electrodes and Montage

Standard silver/silver chloride EEG disc electrodes were used for all the surface recordings. Silver chloride coating decreases the polarization when placed on the skin. The dome shape of the electrode helps to insulate from the skin and to prevent direct contact with the skin and unstable DC potentials charges. Electrodes must be maintained in the freshly chlorided state and frequently replaced to minimize electrical noise.

The montage used is derived from the work of Hume and Cant[35]. The cervical electrode is placed on the skin over the second cervical spine with collodion, and the surface recording scalp electrodes placed at the C 3/C 4 positions of the international 10–20 system, 2 cm posterior to vertex and 7 cm from the midline. Reference electrodes were placed at midfrontal (Fp_z) locations beneath the hair line in awake patients to avoid interference from movement of facial muscles, and at the nasion in anaesthetized patients. There was no difference in the wave configuration between these two reference sites.

The electrode-skin contact impedance must be low to obtain maximum current recording. The greasy surface of the skin, together with some cell debris provides a high resistance. Scratching the skin surface is helpful in reducing the size of electrodermal potentials. In this study, the electrode impedance was controlled by scratching the skin with a blunted needle, through a hole in the electrode cup. Conducting jelly made from sodium chloride and glycerol was put into the cup to ensure good electrical contact. The impedance was checked and maintained for each recording between 1.5

and 5 kohm. For intraoperative recordings the impedance was kept below 2 kohm.

In a recent publication Schwerdtfeger and Ludt[87] demonstrated the suitability as electrodes of stainless steel skin staples, which have the advantage of quick and easy application, resistence to mechanical stress and the possibility to leave them in place for a long time.

Stimulation

SEPs have been obtained with transcutaneous electrical stimulation of the median nerves at the wrist by using square waves of 0.15 m duration delivered at a rate of 3–4 per second. The stimulus intensity used was 2 to 3 times the subjective threshold in order to obtain a supramaximal response. An intensity to elicit a small twitch of the thumb was considered to be adequate.

Bipolar stimulating electrodes were used with the anode placed at a site immediately medial to the radial pulse, at the level of the styloid process of the radius, with the cathode 3 cm proximal.

Recording Apparatus

A Digitimer D 200 signal averager system was used with a number of programmed parameter settings which could be manipulated by a keyboard. The responses were amplified 10^4 times using a Digitimer D 150 with a filter of a time constant of 300 msec and a high cut off frequency of 1 kHz. Either 128 or 256 responses were averaged on a Digitimer D 200. A high resolution graphical display incorporated 2 cursors for latency and amplitude measurements. Data could be stored by an integral magnetic floppy disc. Signals could be plotted on any kind of paper by an X-Y digital plotter. The whole system including an oscilloscope, amplifiers and a signal averager was mounted on a mobile trolley which could be used at the bedside or in the operating theatre.

Table 1. *Control Data From 28 Normal Volunteers*

Stimulation site	N 14	N 20	CCT
Left median nerve (n = 28)	13.5 ± 0.8	19.0 ± 0.9	5.4 ± 0.4
Right median nerve (n = 28)	13.6 ± 0.8	19.1 ± 0.9	5.5 ± 0.4
Overall average (n = 56)	13.5 ± 0.8	19.0 ± 1.0	5.4 ± 0.4
IHD (n = 28)	0.2 ± 0.2 ($p < 0.001$, paired t-test)		

Values are means ± standard deviations (msec).
IHD, interhemispheric difference.

It is important that a service planning to use this technique obtains normal controls of its own, and most laboratories reporting the studies have done so with remarkable consistency between one laboratory and the other. Our own conduction time measurements from 28 normal subjects, 20 males and 8 females, gave results very similar to the initial reports from Hume and

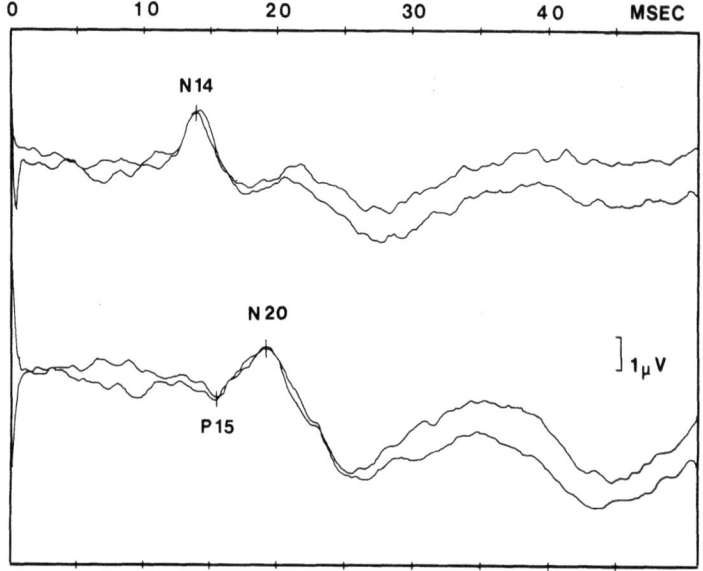

Fig. 1. SEP recording of a normal subject. The top traces were recorded from the second cervical spine (N_{14} peak); the lower traces were recorded from the contralateral scalp (N_{20} peak) following median nerve stimulation at the wrist. The N_{14}–N_{20} interpeak latency is defined as central conduction time (CCT; Hume and Cant 1978)

Cant[35] in New Zealand, CCTs of 5.4 ± 0.4 being recorded by ourselves and 5.6 ± 0.5 being recorded by Hume and Cant (Table 1 and Fig. 1). There is no great variation in conduction time with the age. Hume and coworkers[38] showed an increase in conduction time by 0.3 msec after 50 years of age but we have felt in unnecessary to correct for statistical purposes with advancing age and Desmedt[19] denies significant difference in conduction time between subjects of varying ages.

*The Measurement of SSEP a Prognostic Guide
in Subarachnoid Hemorrhage*

The application of evoked potential analysis to the study of cerebral vascular disease was initially disappointing, probably because of the multilevel involvement of various pathways which rendered attribution of specific levels of ischemia to disorders of the pathway virtually impossible. In subarachnoid hemorrhage, however, where global cerebral perfusion has been shown to be affected [21, 27, 41, 55, 56, 72] it has been possible in many cases, to relate conduction time to, for example, the admission grade using the unmodified scale of Hunt and Hess [39, 83]. An early study carried out on the patients admission, invariably within two weeks of haemorrhage, was followed by recordings throughout the postoperative period, on the first postoperative day, between 48 and 72 hours, at 5 days and then again at 2 weeks. Pre- and postoperative conduction times have been recorded from approaching 200 patients with aneurysmal subarachnoid hemorrhage, and data from 117 cases was reported [71, 83] and is summarized in Table 2. In all cases direct intracranial surgery was carried out, and the affected hemisphere of the table is either the one responsible for any neurological deficit or the aneurysm bearing hemisphere, or in midline cases, anterior communicating and basilar arteries, the side from which the operation was carried out.

In the preoperative period, all patients in grade 2, 3, or 4 showed a significant prolongation in conduction time in the affected hemisphere compared to the normal data. In the unaffected hemisphere conduction time was prolonged with a high significance ($p < 0.005$) in grade 3 or 4 patients. Prolongation of conduction time bilaterally was also pointed out by Fox and Williams [22]. Intergrade differences, however, could only be detected between grades 4 and the other grades, there being no detectable statistically significant prolongation of CCT between grades 1 and 3, in the affected or the unaffected hemisphere. Following the conduction times over the postoperative period showed a slight but significant prolongation following direct intracranial aneurysm surgery ($p < 0.005$) from 5.7 ± 0.4 msec to $6.1 + 0.6$ msec, without necessary clinical deterioration in patients in grade 1. The slightly prolonged conduction times in the other grades when present preoperatively showed no detectable change in the postoperative period unless there was the development of a clinical neurological deficit. In the same way, a detectable change in conduction between the two hemispheres was demonstrable in grade 1 patients postoperatively who invariably showed a slight increase following surgery. In the preoperative period there was no significant widening of the interhemispheric difference except in grade 4 patients.

The significance of these slight prolongations in conduction time have been most recently analyzed [65] in terms of cerebral blood flow. Taking the level of 6.2 msec as two standard deviations above our normal conduction time, it has become apparent that hemispheral blood flow determined by the xenon clearance technique is associated with normal conduction time down to levels of blood flow of 30 ml/100 gm/min. Below that level there appears a

Table 2. *CCT and IHD in Different Time Interval in Relation to Clinical Grades*

Grade		Time interval				
		Preoperative	1 day	48–72 hours	5 days	2 weeks
1	Affected hemisphere	5.7 ± 0.4 (n = 14)	6.1 ± 0.6** (n = 15)	6.0 ± 0.6 (n = 12)	6.0 ± 0.6 (n = 13)	5.8 ± 0.6 (n = 15)
	Unaffected hemisphere	5.7 ± 0.5 (n = 14)	5.9 ± 0.6 (n = 15)	5.9 ± 0.4 (n = 12)	5.9 ± 0.5 (n = 13)	5.8 ± 0.4 (n = 15)
	IHD	0.1 ± 0.1 (n = 14)	0.3 ± 0.3* (n = 15)	0.3 ± 0.3** (n = 12)	0.2 ± 0.2 (n = 13)	0.2 ± 0.3 (n = 15)
2	Affected hemisphere	5.8 ± 0.4 (n = 42)	5.9 ± 0.4 (n = 39)	5.9 ± 0.4 (n = 37) 2 flat	5.9 ± 0.5 (n = 32) 2 flat	5.8 ± 0.5 (n = 38)
	Unaffected hemisphere	5.6 ± 0.3 (n = 42)	5.7 ± 0.4 (n = 39)	5.7 ± 0.4 (n = 39)	5.6 ± 0.4 (n = 32)	5.6 ± 0.4 (n = 38)
	IHD	0.2 ± 0.3 (n = 42)	0.2 ± 0.3 (n = 39)	0.3 ± 0.2** (n = 37)	0.3 ± 0.3 (n = 32)	0.2 ± 0.3 (n = 38)
3	Affected hemisphere	5.9 ± 0.4 (n = 35)	6.0 ± 0.5 (n = 35)	6.0 ± 0.5 (n = 34)	5.9 ± 0.4 (n = 31)	5.9 ± 0.4 (n = 31)
	Unaffected hemisphere	5.7 ± 0.4 (n = 35)	5.8 ± 0.4 (n = 35)	5.8 ± 0.4 (n = 34)	5.8 ± 0.4 (n = 31)	5.7 ± 0.3 (n = 31)
	IHD	0.2 ± 0.3 (n = 35)	0.3 ± 0.4 (n = 35)	0.3 ± 0.3 (n = 34)	0.2 ± 0.2 (n = 31)	0.2 ± 0.2 (n = 31)
4	Affected hemisphere	6.5 ± 0.9 (n = 23)	6.9 ± 1.6 (n = 18) 3 flat	6.7 ± 1.6 (n = 17) 3 flat	6.3 ± 1.3 (n = 21) 1 flat 1 died	6.7 ± 1.5 (n = 20) 1 flat 1 died

Unaffected hemisphere	6.2 ± 0.9 (n = 23)	6.5 ± 1.1 (n = 18)	6.3 ± 1.0 (n = 17)	5.9 ± 0.5 (n = 21)	6.2 ± 1.0 (n = 20)
IHD	0.4 ± 0.4 (n = 23)	0.5 ± 1.0 (n = 18)	0.5 ± 1.2 (n = 17)	0.5 ± 1.0 (n = 21)	0.6 ± 1.2 (n = 20)

* $p < 0.025$.
** $p < 0.05$ (compared to preoperative data; unpaired t-test).

Table 3. *CCT and IHD in Different Time Interval in Relation to 2 Months Outcome*

Outcome		Time interval				
		Preoperative	1 day	48–72 hours	5 days	2 weeks
Good	Affected hemisphere	5.8 ± 0.4 (n = 90)	6.0 ± 0.5 (n = 88)	5.9 ± 0.4 (n = 82)	5.8 ± 0.5 (n = 77)	5.8 ± 0.4 (n = 84)
	Unaffected hemisphere	5.7 ± 0.4 (n = 90)	5.8 ± 0.4 (n = 88)	5.8 ± 0.4 (n = 82)	5.7 ± 0.4 (n = 77)	5.7 ± 0.4 (n = 84)
	IHD	0.2 ± 0.2 (n = 90)	0.2 ± 0.3 (n = 88)	0.2 ± 0.3 (n = 82)	0.2 ± 0.3 (n = 77)	0.1 ± 0.2 (n = 84)
Poor	Affected hemisphere	6.5 ± 0.8 (n = 24)	7.0 ± 1.6 (n = 19)	6.9 ± 1.5 (n = 17)	6.6 ± 1.3 (n = 21)	7.0 ± 1.4 (n = 19)
	Unaffected hemisphere	6.1 ± 0.8 (n = 24)	6.4 ± 1.1 (n = 22)	6.2 ± 1.0 (n = 21)	6.0 ± 0.5 (n = 23)	6.3 ± 1.0 (n = 22)
	IHD	0.4 ± 0.5 (n = 24)	0.5 ± 1.0 (n = 19)	0.6 ± 1.2 (n = 17)	0.6 ± 1.0 (n = 21)	0.6 ± 1.1 (n = 19)

significant prolongation in conduction time (Fig. 2). The difference between the level of blood flow apparently to embarrass electrical function, and the other lower thresholds established in experimental ischemia in the primate, probably indicates more the extensive averaging of flows over a wide area of the hemisphere from a relatively uncollimated probe than a true elevation of thresholds.

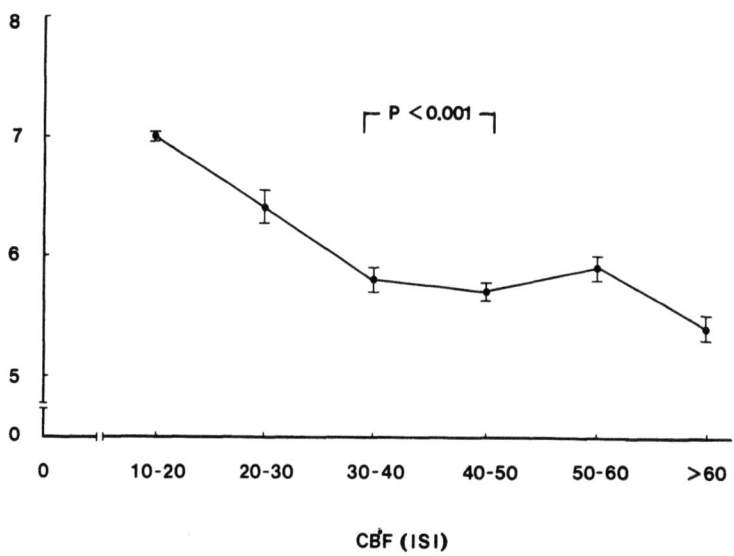

Fig. 2. Relationship between central conduction time (CCT) and initial-slope index of the cerebral blood flow (CBF_{isi}) measured in 120 hemispheres. A threshold phenomenon is noted at flow values below 30 where CCT becomes significantly prolonged. SEM = standard error of the mean (from Rosenstein et al., J. Neurosurg. 62, 1985 with permission)

The Relationship Between Conduction Time, Prolongation and Clinical Deterioration

It is generally true that disappearance of the evoked response from a hemisphere, with the production of an electrically silent trace, occurs only in association with serious neurological deterioration, and a flat trace is not usual in patients who subsequently do well. With frequent recording, the technique was occasionally a premonitor of developing ischemia, but only provided sufficiently frequent recordings were made; the happy coincidence of recording just before the development of an ischaemic episode would reveal slowing in conduction before the appearance of clinical deficit. Regrettably this in clinical terms is less than useful because clinical deterioration might occur quite independent of incidental recording of the conduction time. Conduction time might be normal some hours before the development, for example of vasospasm, and only if measurements were being made close to the fairly clinical change, was the electrical activity a premonitor.

The relationship between electrical activity and outcome, however, was clear. An initial report by Wang et al.[83] divided patients into a good outcome group, 91 patients returning to full activity with no or minimal neurological deficit, and a poor outcome group comprised of 26 patients unable to work, requiring long-term care or dying as a result of the hemorrhage or surgery. From this material, Table 3 summarizes the data of conduction time according to the outcome at 2 months showing a significant difference between the two groups at any interval of time. Interhemispheric difference was also of utility, significant prolongation being associated with a poor outcome from the first day onwards. The development of prolongation of conduction time in the postoperative period, therefore, can be a useful prognostic indicator of outcome of surgery. Such a relationship between clinical deterioration and prolongation of conduction time has also been reported by Kidooka et al.[48].

Monitoring electrical conduction during the recovery phase from aneurysm surgery has not yet reached the stage of analysis at which a one to one relationship between electrical change and clinical deterioration can be established. Grouping data is of use in appraisal of a series of cases but in the direct guidance of one case may occasionally be misleading. However, the development of prolongation of conduction time should alert the surgeon to the possibility of complications, and the illustration in Fig. 3 shows that prompt reversal of ischemic deficit by hyperperfusion, happily signalled in this case some hours before by the development of prolonged conduction time, can be associated with the remarkable resolution of symptoms. In this case hypervolemia and Metaraminol infusion resulted in elevation of rCBF and resolution of hemiparesis.

Rather more disturbing clinically is the occasional small but dense ischemic lesion which may appear with no change in somatosensory conduction. Thus, for example microembolization or operative occlusion of a small opercular branch of the middle cerebral may lead to significant dysphasic defect in the presence of normal sensory function and this has been observed on rare occasions. As in every neurological analytical technique, the integrity of the whole pathway has to be considered. The experimental work of Ladds and her associates[51] has shown that deep lesions will substantially alter the relationships between blood flow and cortical electrical activity, thus thalamic ischemia will result in the disappearance of cortical electrical activity despite normal cortical blood flow. Only by insight and appreciation of the physiological principles involved can management error be avoided in these circumstances.

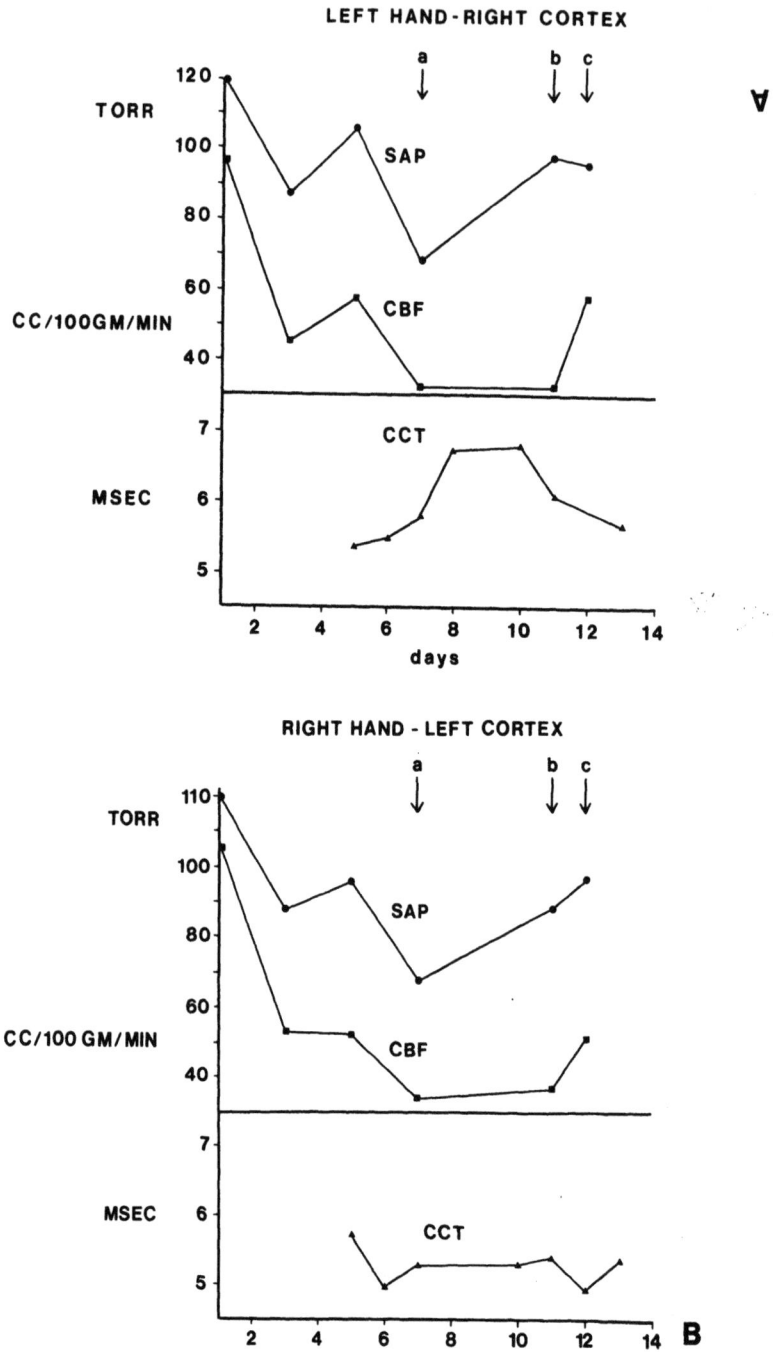

Fig. 3. Serial changes in systemic arterial pressure (SAP), hemispheral blood flow (CBF), and central conduction time (CCT) from a patient who showed prompt recovery of ischemic deficit are shown. In this case left hemiparesis was completely resolved by hypervolemia and Metaraminol infusion, associated with the elevation of rCBF and SAP. Left median nerve stimulation showed significant correlation between changes in CCT and neurological signs (A). Right stimulation showed no significant change in CCT during the ischemic episode (B). *a* onset of left hemiparesis, *b* resolving left hemiparesis, *c* hemiparesis cleared

The Use of Somatosensory Evoked Responses as a Monitor During Intracranial Aneurysm Surgery

The advent of the operating microscope enormously increased the potential for direct surgery on intracranial aneurysm. The major problems of brain swelling and vasospasm have been increasingly analyzed in relation to the ideal timing of surgical approach but there is no doubt that the relatively small exposures required during the use of the microscope have considerably eased surgery in the phase of brain swelling, and in suitable, that is good grade cases, early surgery is now routinely practised in many clinics throughout the world.

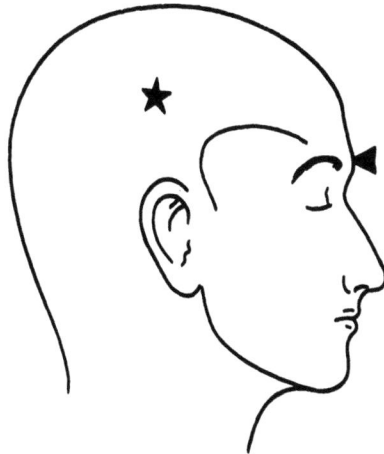

Fig. 4. The site of an aneurysmal flap for an anterior cerebral aneurysm. The approximate site of recording electrodes for SEP is shown by the star, the reference electrodes Fpz by the triangle

The common surgical problem of premature rupture of the aneurysm and the necessity to secure a bloodless field remains. Over the years a variety of techniques have been employed to aid these objectives, varying from deep hypotension to hypothermia with cardiac arrest. Most recently, there has been a more general acceptance of transient proximal vessel occlusion as an adjunct to aneurysm dissection and in this regard the continuous monitoring of evoked electrical activity has proved of value.

Techniques

The standard placement of somatosensory evoked response electrodes referred to earlier with an indifferent on FPZ, an active electrode over the somatosensory cortex, and an electrode on the C 2 spine to record the N 14, all leave the common areas of approach to anterior or posterior circle aneurysms free of electrode sites (Fig. 4). Indeed the relatively small commonly used fronto-temporal scalp flap is well clear of all electrodes and provided the electrodes are carefully fixed, their impedance carefully checked beforehand and their position secured by adhesive tape, they may be comfortably included in the drapes and left undisturbed throughout the procedure. Stainless steel skin staples can be used safely as intraoperative electrodes (Schwerdtfeger and Ludt[87]).

The group in Queen Square have recorded somatosensory evoked

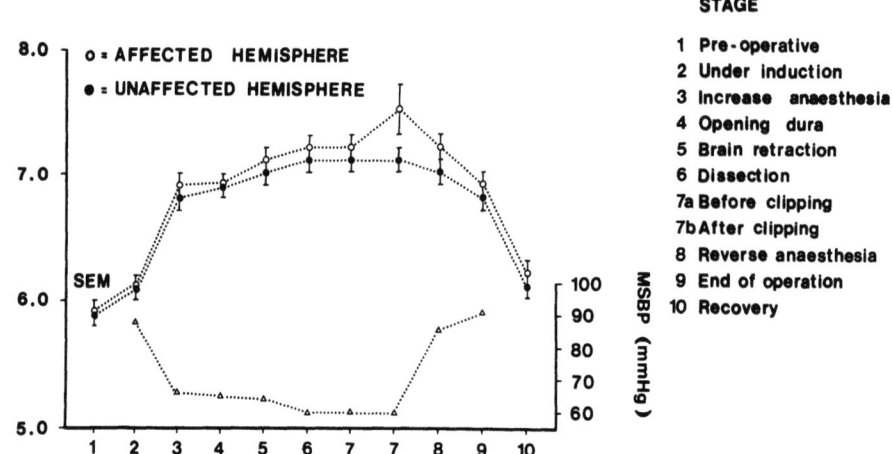

Fig. 5. CCT and MSBP change in the course of surgery. The change was seen in both hemispheres symmetrically except during the phase of clipping of the aneurysm (p 0.001)

responses now from over 100 aneurysm cases and the data from a published series of 68 is shown in Fig. 5.

Major interference with conduction occurs with deepening of anesthesia. The halogenated anesthetics notably halothane produce depression of the amplitude of cortical waves and a gradual increase in latency which is independent of the mild hypotension induced by these agents. Isoflurane is a little less active in depression than halothane but will still, in concentrations of 2%, appreciably depress the somatosensory response. The N 14 neck peak has been fairly constant throughout surgery despite anesthesia and hypotension and of course, the general metabolic disturbance produced by anesthesia is reflected in bilateral changes in SSEP, whereas changes referrable to surgical manipulation are as a rule, related only to one side. This, however, is not necessarily the case in relation to basilar artery aneurysms where temporary occlusion of the basilar artery will occasion disturbance in both SSEP recordings.

Even at low concentrations of halothane (0.5%) all patients show bilateral significant increases in CCT without significant blood pressure change. When halothane concentration reaches more than 1.5%, a further considerable prolongation in conduction time occurs with depression in amplitude of the wave and a significant fall in blood pressure[84].

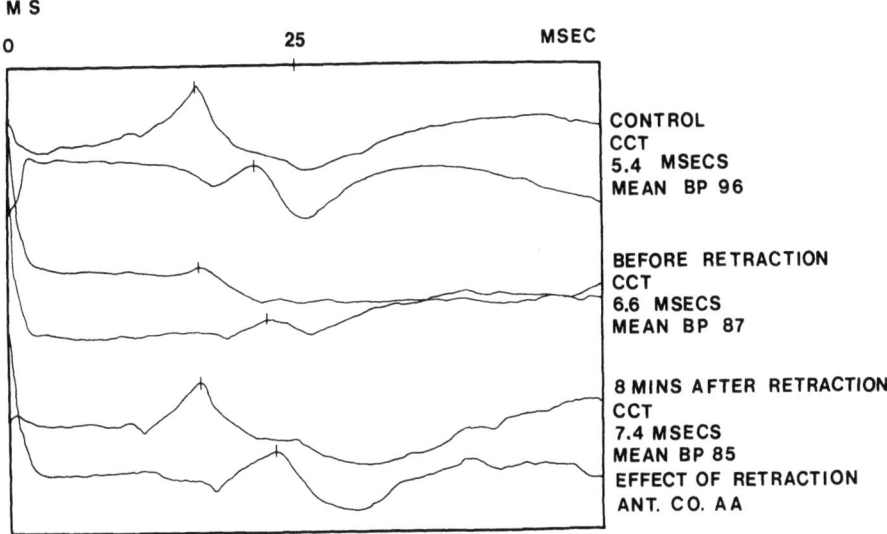

Fig. 6. Effect of brain retraction on CCT. Brain retraction prolonged central conduction time (CCT) in the operated hemisphere from 6.6 to 7.4 msec (from Symon et al., J. Neurosurg. 60, 1984 with permission)

Brain retraction may have a conspicuous effect on conduction time if the retractor pressure is too high. Regional blood flow has been shown to be reduced as a result of retractor pressure especially when the mean systemic arterial pressure is lowered[75] and although in the majority of our over 100 cases, the application of self-retaining retraction was without affect, in 4 cases of middle cerebral aneurysm, 1 each of anterior communicating, posterior communicating, terminal carotid, and basilar aneurysm, application of self-retaining retraction increased the CCT by from 0.4–1.8 msec in the retracted hemisphere. All but one case returned to the preretracted CCT level after adjustment of the retractors. The location of the aneurysms is probably without significance, although the close proximity of the retractor to the somatosensory cortex in the case of middle cerebral aneurysms may explain the relative frequency of this disturbance in the case of this particular location. An example of this type of delay is shown in Fig. 6.

SSEP Recording as a Monitor During Temporary Vascular Occlusion

In over 50 cases of intracranial aneurysm temporary vascular occlusion has been applied to major vessels during SSEP monitoring. A series of 40 such cases has recently been reported[58]. This technique has been found of considerable value in aneurysms of the basilar artery, terminal carotid

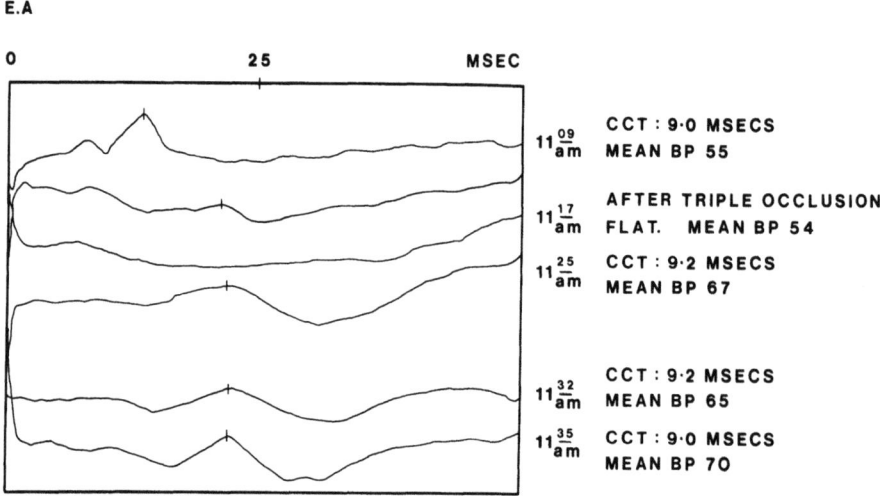

Fig. 7. During surgery, central conduction time (CCT) became flat for 8 minutes after a triple occlusion lasting 1 minute 20 seconds. The upper neck trace (which was constant throughout) and serial tracings from the scalp before the triple occlusion, just before release of the occlusion and 8, 15, and 18 minutes following release, are shown. BP = blood pressure in mm Hg (from Symon *et al.*, J. Neurosurg. 60, 1984 with permission)

artery, and middle cerebral artery. In cases of terminal carotid artery aneurysm 4 of 16 cases with temporary carotid artery occlusion showed prolongation of conduction time immediately after the application of a temporary clip. In the case of the middle cerebral artery, 2 of 12 occlusions showed such prolongation. Conduction was completely lost in 15 cases in this series, 7 being associated with postoperative morbidity of more than transient duration. Considerable prolongation of conduction time to approaching 10 or 11 msec, at which point the detection of the wave form may become very difficult, has been seen in three other cases, one with clips on the carotid, proximal and distal to a large aneurysm at the communicating origin and a clip to the posterior communicating artery itself (Fig. 7), and one giant middle cerebral aneurysm with clips on proximal middle cerebral and distal opercular branches. The third case was a basilar tip aneurysm with clips on both posterior cerebrals and the basilar artery. It proved difficult precisely to relate either the duration of temporary vascular occlusion or the length of time which the wave forms took to recover, to subsequent morbidity, but it was soon clear that without gross prolongation of the N_{20} wave or the attainment of a flat trace, temporary vascular occlusion appeared to be safe in relation to this group of aneurysms. No case with occlusion of the terminal carotid artery of less than 12 minutes or occlusion of the middle cerebral artery of less than 16 minutes has resulted in permanent neurological deficit. No case with such a brief occlusion has

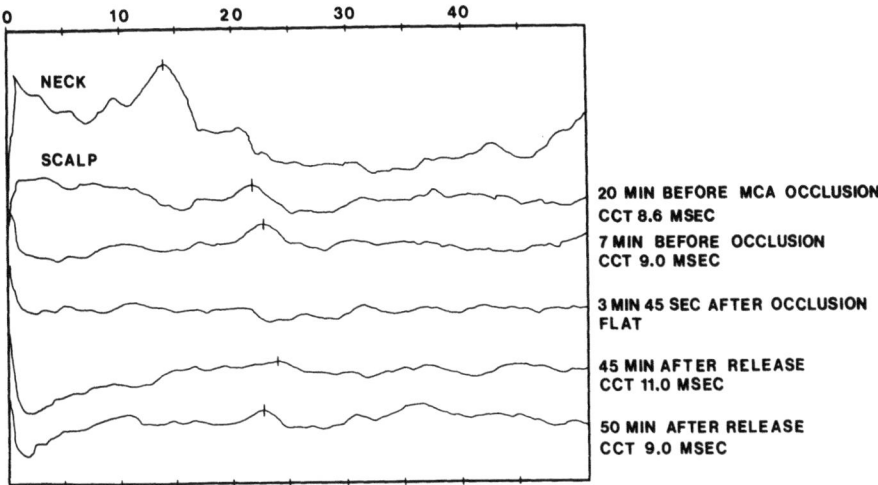

Fig. 8. The upper neck trace and serial tracings from the scalp are shown. The cortical potential became flat 3 minutes and 45 seconds after the application of proximal and distal middle cerebral clips. Duration of occlusion was 6 minutes and 18 seconds. The N_{20} peak took 45 minutes to recover after release of the clips

been associated with any more than transient neurological deficit usually resolving within 48 hours postsurgery. A flat trace has been sustained in one case for 45 minutes with gradual recovery by the end of the operation and no permanent neurological deficit. The shortest period of time with which a complete loss of conduction during operation has been associated with postoperative death was 67 minutes, and this was a terminal carotid giant aneurysm in which attempted ligature of the aneurysm neck at the ophthalmic origin disrupted the internal carotid artery. The attainment of a flat trace relatively slowly after first occlusions, was attained very much more rapidly when occlusions were reapplied in an attempt to repair the neck, and finally ligation of the carotid in the neck and the head resulted in hemispheral infarction.

Of greater utility in terms of prognosis, as far as the surgeon is concerned, has been rapidity of disappearance of the trace. Thus, in 15 cases in which flat traces have been obtained, only 7 in which the trace became flat within 4 minutes of the application of temporary vascular occlusion were associated with appreciable neurological deficit. One of the fifteen, a giant middle cerebral aneurysm in which the perforating branches of the middle cerebral arose close to the aneurysm neck, showed disappearance of the trace in just over 3 minutes following the application of proximal and distal middle cerebral clips. Rapid evacuation of the aneurysm with the ultrasonic dissection apparatus enabled a neck to be fashioned and a permanent clip applied in 6 minutes and 30 seconds, but even so the patient awoke with a dense hemiplegia, extending upper and lower limbs and completely aphasic. Within 4 hours the hemiplegia had virtually resolved, by that evening he was once again speaking normally, and within 48 hours the deficit had cleared completely (Fig. 8). There can be little doubt, however, that this closely approximated to the maximum time of occlusion permissible under these circumstances, reflecting the fact that the collateral supply to the perforating segment is poor and that induction of ischemia in the basal ganglia and capsule will result in blood flow of less than 10% of basal, well below the thresholds both for ionic movement and for electrical failure. In the same way, the rapid disappearance of the trace with proximal basilar and posterior cerebral occlusions in a further giant aneurysm, was a premonitor of intense ischemia in both thalami, which in this instance resulted in progressive infarction and the death of the patient.

Very occasionally conduction time may alert the surgeon to the misapplication of a clip. In 100 consecutive cases, 2 cases of middle cerebral artery aneurysm were experienced in which an apparently perfect clip placement resulted in appreciable prolongation of conduction time and depression of amplitude, resulting in a more careful inspection of the clip placement and the indication in both cases that a posteriorly placed major middle cerebral branch had been partially occluded by what appeared at first a perfect clip. Repositioning of the clip restored the situation in both circumstances with immediate recovery of CCT and no postoperative neurological deficit in either case.

During the reversal of anesthesia a slight but definite prolongation of CCT may be observed as PCO_2 undergoes a rise sometimes into the low 50s with reversal of respiration. It is our standard practice to reverse respiration unless there is particular brain swelling before closing the wound, as a check on venous hemostasis. Such prolongation has in our hands been slight, between 0.4–1.2 msec, usually transient and disappears with the resumption of completely normal PCO_2 levels. 12 cases in our first reported series showed such transient prolongation and in all instances the PCO_2 was above 45 mm mercury during the reversal phase. It is at least possible that this bilateral depression of conduction is the effect of hypercapnia.

During manipulation of the basal vessels in relation to aneurysm surgery a slight prolongation of conduction time has been reported[75] although not consistently observed in other series[48]. This appears related possibly to mild traumatic vasospasm in a circulation which may already be critically compromised by the aneurysmal subarachnoid hemorrhage, and should probably alert the surgeon to the necessity for extremely careful handling of vessels already partially vasospastic.

In the case of anterior communicating aneurysms, SSEP has been employed[28, 71] in an endeavour to predict the safety of anterior cerebral ligation. In the authors hands this has not proved particularly helpful, the somatosensory area for the hand stimulation commonly employed being outwith the distribution of the anterior cerebral artery. Where prolongation of conduction time is seen as a result of anterior cerebral occlusion it may very well result from the large contribution of Heubner's artery to the capsular circulation in this particular case, and detection of cortical ischemia from proximal anterior cerebral occlusion to median nerve stimulation is probably unreliable. More promising results are achieved by bilateral leg stimulation of the posterior tibial nerve, but the technique is more complicated and the analysis more difficult since separation of the recordings from the two sides may prove less than satisfactory. In 8 anterior cerebral artery occlusions, none showed changes in conduction time to median nerve stimulation and in 1 case after 3 minutes and 50 seconds of bilateral proximal anterior cerebral occlusion without change, the patient developed bifrontal infarction and died in 10 days. The hypothalamic and distal anterior cerebral distribution infarct was not predicted by conduction time measurements from median nerve stimulation.

Somatosensory Evoked Potential Recording and Use of Somatosensory Evoked Responses in Head Injury Management

The pioneering work of the Greenberg group[25, 26] and of Hume and Cant[36] in 1981 established the analysis of somatosensory evoked responses as a useful predictor of outcome in head injuries. Since then a number of groups have used the method[52, 54, 61] and the experience here described has been collected from the Department of Neurosurgery, the Saarland University Medical School in Homburg, based upon a group of 42 severe head injuries whose age distribution, incidence of intracranial bleeding and outcome was fully compatible with other major head injury studies[6, 23, 43, 63]. The method of recording of somatosensory evoked responses in similar to that described in relation to the monitoring of ischemic processes, but some groups have found that in addition to the midfrontal reference, further information may be obtained using a contralateral ear reference, noting that P 15 (the subthalamic or brain-stem component) tends to be larger with the ear reference.

As before, electrode impedance must be low (below 5 k ohm) and the amplified band pass with in the Homburg study was 1.5–3 kHz. Once again, successive series of 256 or 512 responses with a post stimulus interval of 200 msec were averaged.

The principal measurement used in relation to head injury management has again been the central conduction time[36] but it may be worthwhile in

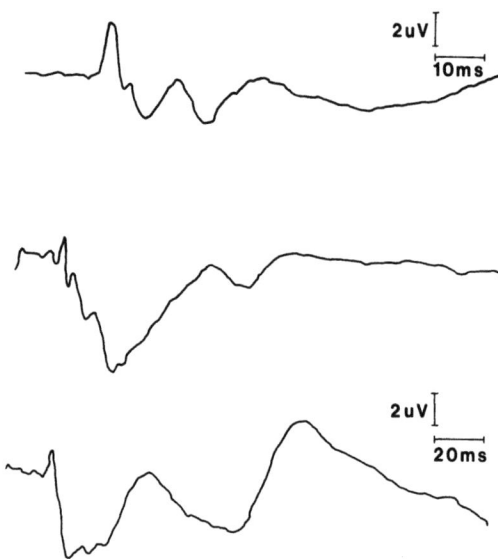

Fig. 9. The reliability of the SEP waveform 3 examples of normal subjects. Note different time calibration in the upper row showing the first cortical answers in greater detail

head injuries assessment to identify 7 peaks in the 200 msec poststimulus interval, the number fairly constantly found being somewhat less than 11 theoretically possible. Complexities of the N 20-P 26 complex[35] renders the analysis of these later waves somewhat difficult and indeed some individuals show two additional peaks in the 50 msec latency range and the cognitive dependent component, the N 140 wave which could occasionally be recorded (Fig. 9). These later components tend to be augmented by reduction of stimulus frequency[3] but the sequence pattern evoked at 3 per second is unchanged.

The principal measurement has been central conduction time which in 11 controls in the Homburg series was 5.6 ± 0.46 msec with a maximum interhemispheric difference of 0.5 msec. The grading system suggested by Greenberg[26] was applied to the cortical response, consisting of a four-grade scale depicted in Table 4.

Table 4. *The Greenberg Scheme: Graded Injury Potentials in Head Injury*

SEP Wave* Grade:	I	II	III	IV
P 15	+	+	+	+
N 20	+	+	+	−
P 26	+	+	−	−
N 36	+	+	−	−
P 53	+	+/−**	−	−
N 91	+	−	−	−
P 128	+	−	−	−

* Adapted to our normative data.
** Occasionally present.

Clinical evaluation of patients in such head injury study groups is generally made according to the Glasgow coma scale[78] (Table 5) including tests of pupillary reflex and motor reactions. Standardized treatment schedules will include early surgery for space occupying lesions, anticonvulsant medication (phenobarbitone 400 mg/day in the Homburg series). Mannitol infusions as necessary and management of concurrent injuries. Systemic blood pressure has generally been kept controlled between 100 and 140 mm of mercury and controlled artifical respiration instituted with PCO_2 and PO_2 levels maintained at 25–35 mms of mercury and above 100 mms of mercury respectively. Steroid treatment has not been universally applied. Intracranial measurement has not been systematically measured and CSF has not been drained.

Any infection has been treated with antibiotics and elevated body temperature lowered but hypothermia has not been employed.

Statistical evaluation of averaged signals has used Fischer's exact probability test. Contingency of distributions was evaluated with Pearson's coefficient of contingency as percentage of the maximal coefficient for four field tables[66].

The Homburg group have assessed outcome, one or two years after trauma on the five point scale described by Jennett and Bond 1975[42] (Table 5).

Clinical Factors Related to Outcome

The following factors established at admission related to outcome (within the categories GR/MD and SC/PVS/D); age, GCS score, pupil reaction and best motor response (Table 6). Age was the most significant

Table 5

A. The Glasgow coma scale

Eye opening	spontaneous	E 4
	to sound	3
	to pain	2
	never	1
Motor response	obeys commands	M 6
	localizes pain	5
	normal flexion (withdrawal)	4
	abnormal flexion	3
	extension	2
	nil	1
Verbal response	oriented	V 5
	confused conversation	4
	inappropriate words	3
	incomprehensible sounds	2
	none	1

B. The Glasgow outcome scale

GR	good recovery	resumption of normal life
MD	moderate disability	disabled but independent
SD	severe disability	conscious but dependent
PVS	persistent vegetative state	prolonged unresponsiveness
D	death	

C. Motor function

normal power	5
moderate weakness	4
severe weakness (antigravity)	3
trace movement	2
paralysis	1

Table 6. *Clinical Findings Related to Outcome*

	GR/MD	SD/PVS/D	p*
GCS score			
3–5	2	4	
6–8	5	10	n.s.**
> 8	1	2	
Pupil reaction			
bilateral impaired	1	8	n.s.
normal/unilateral impaired	7	8	
Best motor reaction			
< M 3***	2	3	n.s.
⩾ M 3	6	13	
Age			
< 30	0	11	< 0.005
⩾ 30	8	5	

* Fisher's exact probability test.
** Significance was calculated adding the 6–8 and the > 8-subgroups.
*** Motor reaction expressed in terms of the GCS (*cf.* Table 1).

variable, an age over 30 years implying a poor prognosis (Pearson's contingence coefficient, corrected for 2 × 2 tables equals 61%), as bilateral impaired pupil reaction did, to some degree, too (Pearson's coefficient = 50%). No relationship between outcome and best motor response or GCS score could be detected. Interdependence of the clinical signs at admission was tested. Some degree of dependence was found between a low GCS score and bilateral absent pupillary reactions and between posturing or flaccidity and bilateral unresponsive pupils.

SEP Related to Outcome

We, in common with Hume and Cant found that the central conduction time as already defined was the best predictor quite superior to the Greenberg scheme[25, 26] (Table 7). Some additional power of prediction was revealed by taking into account the age. However, none of the SEP parameters measured, could predict outcome in every single case. A well

Table 7. *SEP Parameters and Outcome*

A. Classification of Greenberg			
a) best response	I/II	III/IV	p*
GR/MD	6	1	n.s.
SD/PVS/D	14	2	
b) worst response			
GR/MD	6	2	n.s.
SD/PVS/D	10	6	
B. Total numbering of waves			
	⩾10	<10	
GR/MD	4	4	
SD/PVS/D	9	7	n.s.
C. CCT			
a) best response	<3 SD (7.0 ms)	⩾3 SD	
GR/MD	7	1	
SD/PVS/D	8	8	n.s.
b) worst response			
GR/MD	4	4	
SD/PVS/D	4	12	n.s.

* Fisher's exact probability test.

Table 8. *Age Matched SEP Data*

Age	<30		⩾30	
SEP*	I/II	III/IV	I/II	III/IV
GR/MD	6	2	0	0
SD/PVS/D	2	3	8	3****
CCT**	<3 SD (7.0 ms)	⩾3 SD	<3 SD	⩾3 SD
GR/MD	7	1***	0	0
SD/PVS/D	2	2	7	5

* Including the worst response.
** Including the best response.
*** See text.
**** See text.

Table 9. *Prognostic Factors for Motor Outcome*

	No paresis (5*)	Paresis (<5)	p
A. *Initial motor reaction*			
⩾M 3	9	11	n.s.
<M 3	4	4	
B. *SEP (Greenberg scheme)*			
I/II	12	8	<0.05
III/IV	1	7	
C CCT			
<3 SD (7.0 ms)	11	3	<0.001
⩾3 SD	1	13	

* See Table 1C.

preserved SEP and/or a normal central conduction time especially when present in both hemispheres, however, indicated a very good prognosis in the younger patient but had less prognostic value in the older one. Even a much distorted SEP did not preclude a reasonable functional recovery in the young. An example is shown in the lower part of Table 8—a 16-year-old girl with initially absent responses over both hemispheres who made an unexpectedly good recovery and is now relearning the skill of writing. In older people, however, a seriously disordered SEP is a prognostically poor sign. Finally, there was a high correlation between CCT values and remaining paresis (Pearson coefficient = 85% Table 9), in line with the assumptions expressed in earlier results of the Homburg group[1,2].

Progressive Change in SEPs

As described by Hume and Cant, progressive changes in somatosensory evoked response morphology could be followed in 24 of 42 cases (57%) during the first posttraumatic weeks. Change took the form of improvement in 12, deterioration in 12, and no change in 4. The remainder either died prior to control records or had to be referred to another institution. 5 of the 12 patients with deteriorating EP patterns made a good or moderate recovery and only 5 of the 12 showing improvement had a favorable outcome so that prognosis could not be inferred from these changes. Table 10 demonstrates the results obtained in a group of later posttraumatic cases between 6 months and 2½ years following accident. Most patients of these groups had a good outcome as shown by a trend to further

Table 10. *SEP at Outcome*

Patient		Sch. N.	L. J.	Sch. K.	Sch. M.	H. S.	M. H.	B. N.
GCS		14	7	6	7	6	7	6
Outcome		MD	GR	MD	GR	GR	SD	PVS
Motor react.	le	M 1	M 5	M 3	M 5	M 2	M 5	M 4
	ri	M 6	M 5	M 4	M 5	M 4	M 5	M 4
Motor force	le	2	5	4	5	5	5	3
	ri	5	5	3	5	5	3	1
SEP (init.)	ri	IV	II	II	I	I	I	I
	le	I	II	I	I	I	I	I
SEP (outc.)	ri	IV	II	I	I	I	I	II
	le	I	I	I	I	I	I	IV
CCT (init.)	ri	no	5.8	7.0	6.9	6.6	5.8	7.8
	le	6.4	5.7	5.1	5.6	6.9	7.5	13.8
CCT (outc.)	ri	no	4.1	6.6	5.9	4.7	6.6	9.2
	le	6.0	4.3	5.6	5.9	5.3	6.6	no

Explanation for the abbreviations of motor reaction, the motor force and the scores of the Glasgow coma scale—see Table 5.

improvement of electrophysiological data compared with the initial value. Exceptions did occur, however, for example a gunshot wound of the right parietal region involving the right central region and parts of the right frontal lobe with gross leftsided hemiparesis, showing poor electrical recordings of grade 4 on the right side but a very considerable improvement to capacity to manage his daily life with only minor mental deficits. A further exception was one of two cases with poor outcome in which despite good preservation of the SEP in a gunshot injury with complete bilateral destruction of the frontal lobes and optic nerves poor outcome classification had to be based upon consecutive blindness and a marked frontal lobe syndrome and not upon the motor power which was quite well preserved.

As referred to elsewhere in this chapter, brain-stem and visual evoked responses have also been employed in head injury but it has been the experience of this group and of others that while evaluation of the outcome of head injury patients with the aid of SEPs shows a clear prognostic value when related to age, the prognostic value of the other modalities is stronlgy dependent on clinical characteristics and generalizations are impossible. In regard to SEP, however, even severe distortion of wave forms does not preclude a favorable outcome in the young, although these cases are rare. In patients over 30 years of age in the same situation prognosis is consistently poor. In this group even an initially well preserved SEP does not exclude an unfavorable outcome.

The highest correlation of SEP with functional failure is with the best motor response and the predictor value of SEP in this case is highly significant. The incidence and quality of intervening complications plays a decisive but as yet not readily quantifiable role in the predictive properties of SEP in head injury.

2. Visual Evoked Response Monitoring

Since the initial description of evoked responses to flash stimulation by Calvet, Cathala, Hirsch, Scherer 1956[13] visual evoked responses have gained a firm place particularly in the analysis of pathway defects secondary to demyelinization. It has become generally accepted that pattern evoked visual responses are more effective than flash visual evoked responses, but the simplicity of the technique of flash VEPs renders them particularly suitable for preoperative use. From the clinical point of view the major drawback is that flash responses are very variable in form when comparing different subjects[50] but by the use of specially modified contact lenses[14, 20, 85] or light emiting diodes mounted on spectacles[14, 50, 64] quite reliable results have been obtained in comatose patients or others with poor cooperation. Jacobson et al.[40] pointed out that the flash VEP stimulations of each eye individually were in any one subject similar in wave from an latency when the responses were recorded form homologous locations over the two hemispheres.

At Queen Square, Halliday et al.[30] reported a series of 19 cases of compressive lesions of the anterior visual pathway studied by means of pattern evoked responses before and after operation. Tumours arising in the region of the sella turcica were associated with a high incidence of abnormalities in the wave form of the response and asymmetry in the field of the occipital evoked potential was especially characteristic in this group. Most but not all the asymmetric cases were associated with a field defect. The incidence of delayed responses was much lower and the magnitude of the delays was smaller in Halliday's experience than those found in the already considerable group of cases with primary demyelinating disease.

In the National Hospital group, a consecutive series of 40 patients with compressive lesions of the anterior visual pathway was studied over a twenty-month period, patients being examined clinically and flash VEPs recorded before and after operation. In 20 cases successful VEP recording was also performed during surgery on the anterior visual pathways. The series consisted of patients presenting with visual deterioration and all were cases of tumor compression of the anterior pathway.

As in other forms of evoked response testing, it is important that control subjects be examined in the same laboratory, and a group of 20 healthy subjects formed the basis of the control group in the operative series studied.

3. Methodology

Chlorided silver cup electrodes 9 mm in diameter were used for scalp recording attached to the scalp with collodion, preparatory cleaning with acetone as usual being performed.

A monopolar (common reference) montage was used, consisting of a transverse chain of three active electrodes 5 cm above the inion one placed in the midline and the other two 5 cm to either side. All electrodes were referred to a midfrontal reference electrode 12 cm above the nasion. These electrode placements are simple in measurement, and may be maintained throughout the performance of standard frontal craniotomy. The other electrode was placed between the frontal reference and the midoccipital electrode. Only the midoccipital midfrontal derivation was used during intraoperative recordings. As with SSEPs, a small square of adhesive was used to cover the electrodes during preoperative recording and saline jelly was inserted into the electrode to provide a low impedance conducting bridge between the electrodes and the scalp. Impedance of the electrodes was always checked and kept to less than 5 kohm. Electrode leads were kept close together by twisting them to reduce the effect of magnetic interference. The electrode montage is shown in Fig. 10.

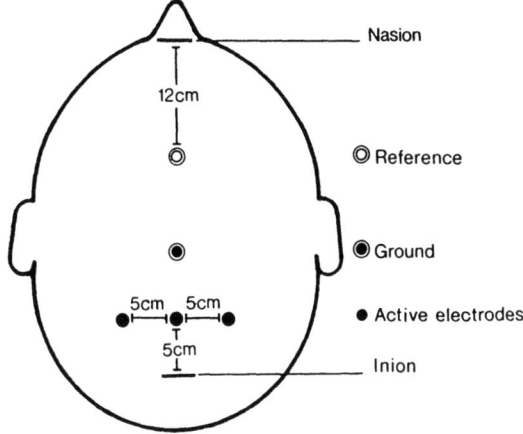

Fig. 10. The montage of surface electrodes. Three silver cup electrodes were placed in such a way that a central electrode was 5 cm above the inion and the other two 5 cm to either side of it. A midfrontal reference was used, placed 12 cm above the nasion

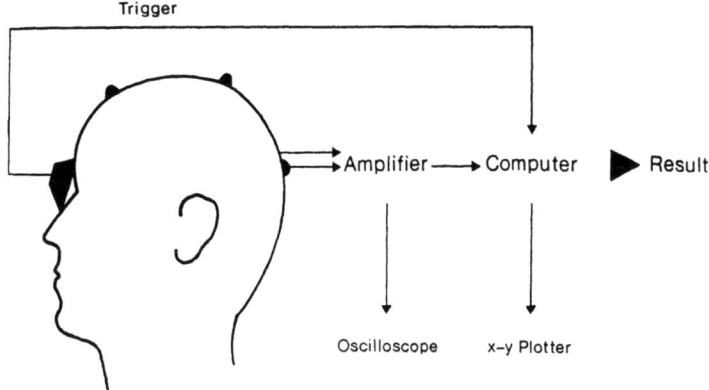

Fig. 11. A schematic diagram of the equipment

The eyes were tested individually using flash stimulation from an array of 10 light emiting diodes (LEDs). These were mounted in a opaque eye patch and during peroperative recording the eyelids were taped closed. The stimulus was delivered at the rate of 2 per second with a peak intensity of 800 candellas for each array. Stimulation in peroperative, operative and postoperative condition was always through the closed eyelids.

As with other systems of evoked response analysis, a wide variety of recording apparatus is now available. However, Queen Square responses were recorded on a Digitimer D 200 signal averager which was also used for SSEP recording. A schematic diagram of the equipment is shown in Fig. 11.

Recording Technique and Wave Identification

As in other forms of evoked response analysis averaging is necessary, and in flash visual VEP analysis 128 responses using an analysis time of 250 msec is generally sufficient. Filter settings gave a band pass from 0.53 to 300 Hz and as a rule three different averages are necessary for each

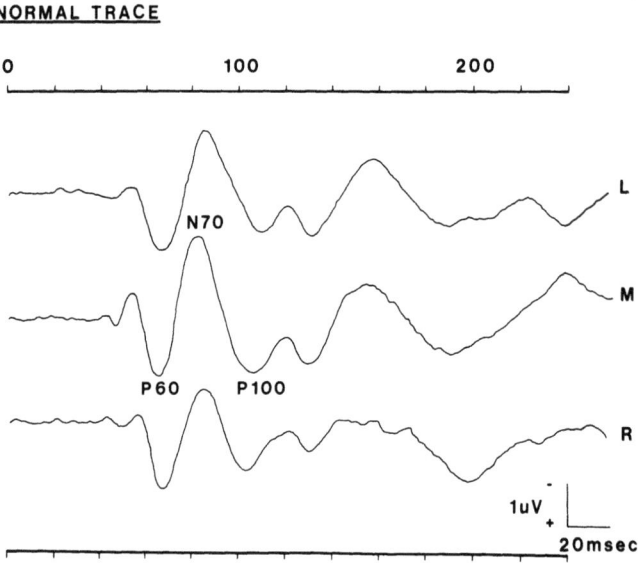

Fig. 12. An example of flash-visual evoked potentials recorded from a normal subject. Special reference is made to the P_{100} peak in this study. (L: left occipital electrode; M: midline electrode; R: right occipital electrode)

condition. If the trace is flat or if there is an ill-defined wave form, averaging without stimulation is necessary to evaluate the contribution of background activity and this is particularly essential where peroparative recording is required.

The standard form of flash VEP is shown in Fig. 12, and the current nomenclature has been used, negative peaks have been recorded upwards positive downwards. The two peaks which can be most readily identified are the N 70 and P 100 which correspond to the wave 3 and 4 of Ciganek 1961[15], or Nz and Pz of Harding 1974[31].

Costa e Silva et al.[17] have described a grading classification of visual evoked responses on the basis of latency amplitude and attempted to correlate this grading with the clinical and surgical characteristics of the patients studied.

VEP as a Guide to Manipulation of the Optic Pathways During Surgery

As in SSEP recording, anaesthesia substantially modifies the visual evoked response, and indeed flash VEPs are extremely sensitive to halogenated anaesthetics. Control investigations were carried out on six patients undergoing anterior cervical fusion[16] after explanation to the patients and the obtaining of informed consent. No significant correlation between prolongation in the P 100 latency and blood pressure change was found, but with increasing halothane concentration there was a significant progressive prolongation in P 100 latency; the two peaks showing most reliable change were the N 70 and P 100 (3 and 4 of Ciganek 1961[15]) the increase in latency of these two peaks was significantly correlated to the level of anesthesia but no evident change in VEP amplitude could be detected as a function of halothane concentration in the expired gas, findings similar to those reported by Uhl[80].

Changes was also noticed when CO_2 concentration was increased for example in the recovery of spontaneous respiration, slight increases in latency being detected under the influence of hypercapnia.

Despite the failure of the wave-amplitude to change in control cases, cases with compression of the anterior visual pathway appeared even more sensitive to halothane concentration and in one of the peroperative cases halothane concentration of 2% completely abolished the flash VEPs. Similar findings were reported by Wright et al.[86] who noted the influence of PO_2, PCO_2 and halothane concentration during the continuous monitoring of visual function in intraorbital surgery.

In 20 cases recorded peroperatively, alterations in the evoked potential were observed in 9 cases as attributable to manipulation of the anterior visual pathway. Operative manipulation of the optic nerve and chiasm is associated with alterations in wave form, in 14 of the recorded cases responses became disorganized and of low amplitude, and in 6 cases there was virtual obliteration of the response.

Operative relief of compression was associated with improvement in the wave form of 55% of our recorded cases and in 20% there was a decrease in latency also. In only 5 cases were changes in the VEP more than transient during operation. The response often began to recover within minutes after the removal of factors associated with transient obliteration such as dissection of the tumour or transient manipulation of the nerve or chiasm. Where this recovery did not take place clinical and FVEP deterioration was found in the first week and a half after operation, and in these cases clinical improvement was more rapid than improvement in the flash evoked response as also reported by Tawana et al.[77]. General experience confirms the view of Wilson et al.[85] that when an optic nerve is manipulated the FVEP will be blocked and when the nerve is decompressed it will be restored. The

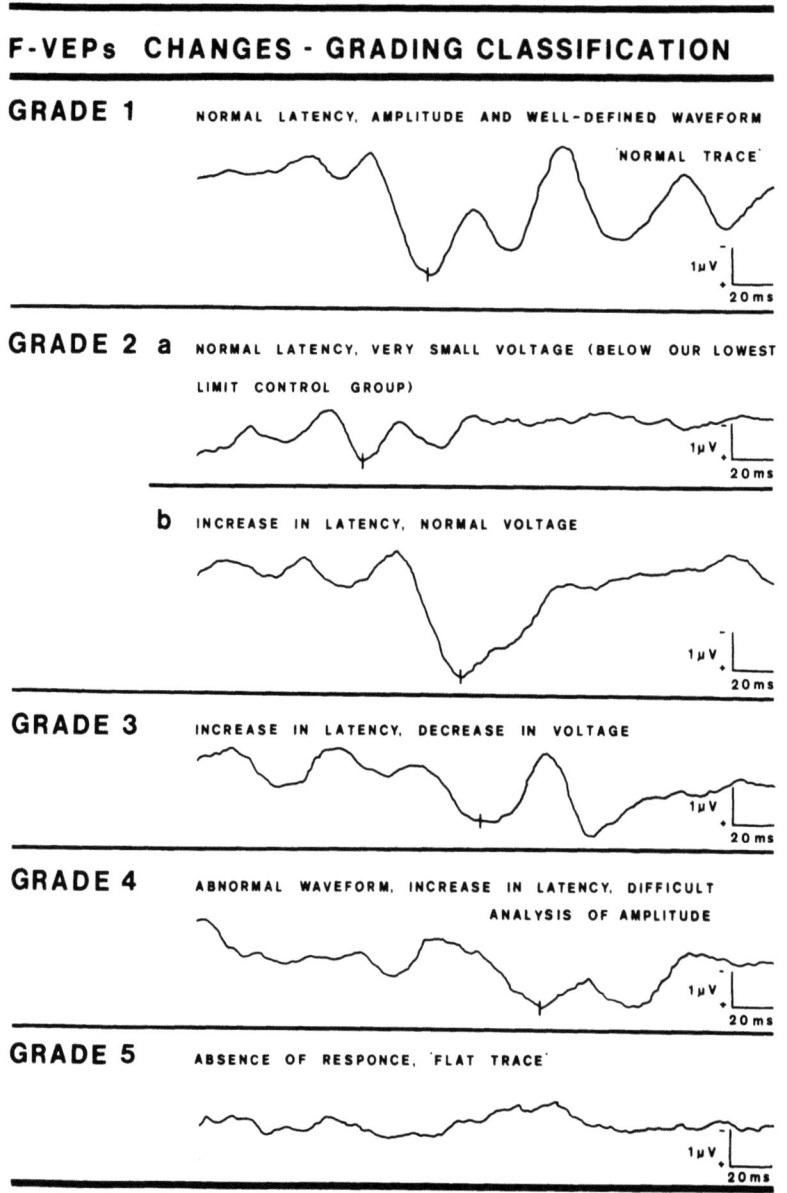

Fig. 13. F-VEP grading classification scheme (from Silva *et al.*, Neurol. Res. 7, 1985 with permission)

explanation is a probable transient abnormality in axonal conduction secondary to ischaemia and the usual time for return of the pattern is 2 to 3 minutes.

As in other series the flash visual evoked response showed a reasonable correlation with preoperative visual defect, in less than 20% of cases of anterior visual compression was the visual evoked response within normal limits and delayed latency was the most reliable criterion. In 8% of our cases flash VEP returned to normal in the first week and 40% of the cases improved to normal vision within the first month. In a further 25% there was appreciable improvement in visual field acuity within the same period of time and in the 16% or so of cases which showed no improvement, the visual phenomena and flash VEP correlated well. Neither improved. The grading system developed by de Silva et al.[17] (Fig. 13) revealed significant differences between visual evoked response, grade and clinical phenomena and indicated also better prognosis following decompression in grade 1 than in grade 5. Tawana et al.[77] studied a series of cases with pituitary tumors showing a decrease in latency of pattern reversal P 100 wave with an increase in its amplitude after operation with change only in acuity. Once again, VEP change correlated well with clinical improvement. Where their cases showed both loss of acuity and field loss, field change was slower in recovery than change in acuity and these findings were repeated in our study.

The flash visual evoked response will follow faithfully the dissection of the optic nerve during operation, 2 of our cases with meningiomatous compression, one from the sphenoidal region, one from the optic nerve sheath, showed irreversible loss of the evoked response—in the case of the sphenoidal ridge following extensive dissection, in the case of the optic nerve sheath meningioma with division of the nerve. After prolonged dissection in the case of this sphenoidal ridge meningioma there was extensive deterioration in the response peroperatively, with slow improvement postoperatively correlated with an initially worse visual function which gradually improved to preoperative status. Factors associated with a favorable outcome include a short history, the use of a direct approach, sparing of the ophthalmic artery and the perichiasmatic blood vessels, and the avoidance of the excessive manipulation of the optic nerve and chiasm with good decompression of the visual apparatus at the time of surgery. These criteria were established by Symon and Jakobowski[74] in an analysis of 101 pituitary tumors over a ten-year period. An example is shown in Fig. 14, in which dramatic improvement of visual function following decompression was assessed by monitoring of FVEP. Rapid visual recovery following surgical decompression has also been well shown by Wilson 1976[85] and Raudzens 1982[64].

Conclusions

The experience of the Queen Square group confirms that of other authors, in that detectable changes in visual evoked response can be correlated with surgical events during chiasmal surgery. Nevertheless, the sensitivity of the responses to the unfavorable environment of the operating theatre with suction apparatus, microscopes, television cameras and in particular diathermy makes their peroperative use exacting. In addition to this, the removal of compressive lesions will necessarily involve some manipulation of the pathway and analysis of VEPs has merely emphasized the necessity for vascular preservation and extreme care during such dissection. It is possible, however, that in the transnasal removal of large

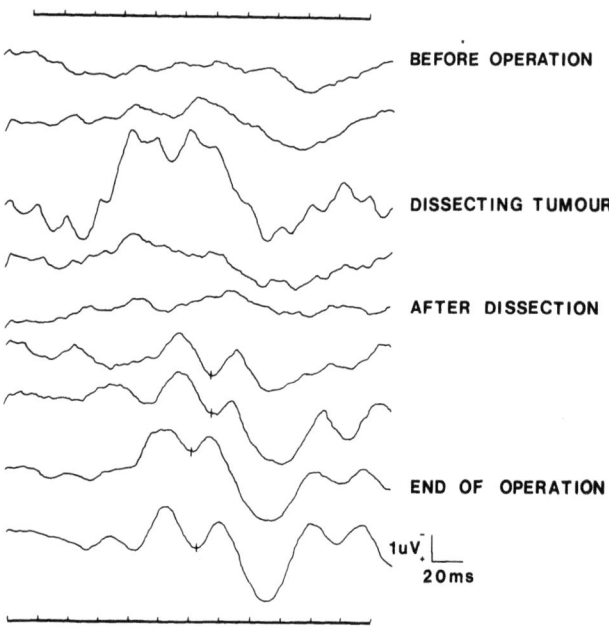

Fig. 14. F-VEP recordings during operation for a pituitary adenoma case. An improved F-VEP waveform was noticed towards the end of operation and the patient had an improved visual acuity and visual fields within three days (from Silva et al., Neurol. Res. 7, 1985 with permission)

pituitary tumors the use of flash VEPs can give early warning of approach to the nerves or chiasm which in the absence of the direct vision available in the transfrontal approach, might avoid excessive blind manipulation of the visual apparatus. As in all other evoked response recording techniques, experience with the technique is essential and each unit must establish its own parameters on the general guide lines described in this chapter.

4. The Use of Brain Stem and Other Auditory Evoked Potentials

Brain Stem Auditory Evoked Potentials in Head Injury

There is an already considerable literature on the use of brain stem auditory evoked responses in head injury. In cases where clinical brain death is suspected the BAEP has been regarded as a useful addition to the analysis of brain stem function [12, 67, 68]. The most optimistic suggestion came from Seals et al.[67] who in a series of 17 patients in coma from blunt head trauma reported a 100% correlation between death and the presence of abnormal brain stem responses. Others, however, have been less optimistic, Goldie et al.[24] being unable to demonstrate the presence of BAEPs in 16% of cases who showed evident continuing brain stem function. Studies by Brewer and Resnik[11] in 52 patients in a surgical intensive care unit, a mixture of blunt head trauma in 38 patients, cerebrovascular disorder in 8, anoxia in 3, gunshots in 2, and encephalitis in 1, showed a reasonable correlation between the brain stem auditory evoked response abnormalities and clinical outcome, a correlation of 77% being abtained in the group as a whole and if deaths from other causes could be excluded an improvement in predictive accuracy to 91%. Accuracy was closest where the Glasgow coma score was poor, between 5 and 7, and where brainstem reflexes were present but abnormal. Even this group, however, in reporting a careful study admitted that BAEP could falsely predict recovery in some patients who expired but never predicted death for patients who survived. On the other hand, Adler et al.[1, 2] found brainstem auditory evoked potentials strongly dependent on clinical characteristics in every single patient and generalization to be extremely difficult. Greenberg's extensive work has suggested that multimodality evoked potential analysis is of use in prognosis in head injuries, but the techniques of pattern recognition are idiosyncratic and have not found general acceptance.

It may, therefore, be stated that at the present time BAEP analysis can be helpful as an adjunct in the determination of brain death although simpler techniques are available, and the opinion of Lindsay[54] that the complexity of the technique adds little to the prognosis in head injury appears to be justified.

Use of Brain Stem Evoked Responses During Posterior Fossa Surgery

BAEPs are easily recorded in the operating room. Silver cup electrodes attached with collodion and filled with conductive jelly are applied to the vertex and ear lobes in the A 1, A 2 position the standard international 10/20 system; our standard reference placement has been on the forehead (FZ). Electrode impedances should be kept to less than 10 kohm. The Homburg group have used stainless steel skin staples also[87]. Acoustic stimuli commonly used are square waves of 0.1 msec duration, in our own series generated by Digitimer D 200 system through ear phones as repeated clicks of alternating polarity at a rate of 9 per second. Stimulus intensity has been 80 decibels, and bilateral stimulation has been used during surgery. BAEP from both sides were simultaneously amplified and filtered, the filter range being between 3 kHz and 100 msec. In general, between 512 and 1024 sweeps to 10 msecs recording are averaged. BAEP recording may be carried out continuously while the patient is anesthetised, placed on the table and through the operating procedure, and it has our standard practice to record the BAEP before operation and postoperatively according to the patient's condition.

We have found BAEP peroperatively to be extremely sensitive to a variety of incidental phenomena. As with the visual evoked response, halogenated anesthetics considerably depress the amplitude and prolong the latency of wave 5 in particular as reported by Jones *et al.*[45] and others[47].

Lumbar drainage in several of our tumor cases showed asymmetrical or symmetrical prolongation of the BAEP and on opening the dura, wave 5 prolongation has been frequently seen. Wave 5 prolongation is also evident unilaterally or bilaterally during brain retraction and in several of our cases a wave 6 appeared during dissection although it was occasionally difficult to differentiate from wave 5.

Despite these clearly incidental and scarcely significant changes, we have found development of an ipsilateral flat trace usually to correlate with considerable adhesion for example of an acoustic neuroma to the brain stem and usually to precede the development of slowing of the pulse or EEG abnormality, the more time honored warnings of transient brain stem disturbance.

It has been general experience including our own that prolonged disappearance of the brain stem evoked responses during surgery indicate postoperative brain stem dysfunction and bilateral wave 5 prolongation would also correlate with slight postoperative brain stem disturbance. 4 of 9 patients recorded in Queen Square showed such correlation between either a flat trace or prolongation of wave 5 and transient brain stem disturbance although in no case was a flat trace associated with permanent neurological deficit, all patients recovering within 36 hours. The most severely disturbed patient showed ipsilateral flat trace with a tumour extremely adherent to the brain stem and considerable peroperative disturbance not only of evoked response but also of EEG and blood pressure. She was extremely drowsy over 36 hours postoperatively with hyperpnoea over this period but thereafter made a complete recovery.

In common with others[82] we were able to detect slight worsening of the BAEP compared with the preoperative findings in some patients to binaural and contralateral stimulation despite perfect postoperative clinical condition, so there is clear evidence of discrepancy of postoperative BAEP and clinical condition.

It seems to us that the sensitivity of the method is perhaps unduly high for routine clinical purposes, the potentials being easily disturbed by insignificant brain stem manipulation. However, it is clear that the BAEP constitutes a most sophisticated early warning system in posterior fossa surgery, and the technique will clearly continue on a research basis.

Electrocochleographic Monitoring During Acoustic Neuroma Surgery

The work of Levine and Ojemann[53, 62] has indicated the utility of direct monitoring of cochlear potentials during surgery of acoustic neuromas. It became clear from our own and work of other groups that the recording of brain stem auditory evoked potentials was of limited value during posterior fossa surgery, since although they could often be correlated with postoperative condition, their sensitivity to relatively minor manipulations made their use as an operative guide somewhat limited.

Using a similar electronic montage, however, supplemented by the use of a transtympanic electrode, it is possible to record the electrocochleogram and the potential from the auditory nerve itself, N 1 of the ECoG or wave 1 of the BAEP. The cochlear microphonics from the hair cells of the inner ear (CM of the ECoG) can also be recorded. We and others have confirmed the pioneering work of Levine and Ojemann and where attempts are being made to preserve hearing in relatively small acoustic neuromas with useful hearing preoperatively, the technique is of extreme utility and can be recommended.

In recent years, we have used electrocochleography in addition to BAEP in surgery of acoustic neuroma where residual hearing persists. Patients have preoperative pure tone audiograms, speech discrimination audiograms and brain stem evoked responses together with electrocochleography to determine the auditory thresholds for eliciting a consistent response. Preoperative baselines are thus obtained for wave 5 of the brain stem potential and the N_1 potential from the electrocochleogram.

The Technique of Peroperative Electrocochleography

Our methods have been adapted from those described by Ojemann and associates[62]. Auditory nerve potential (N_1) and the cochlear microphonic potentials (CM) are recorded by a transtympanic silver ball electrode of 0.5–1 mm in diameter placed through the inferior part of the tympanic membrane under microscopic control at the beginning of the operation. The electrode rests on the promontory of the medial wall of the middle ear cavity and takes as its reference a cup electrode checked for low impedance applied to the ipsilateral earlobe. Standard placements at the scalp and earlobe are used to record BAEP.

A double lumen tube is placed in the external auditory canal and the auditory stimulus transmitted via a 15 cm 12 gauge polyethylene tube into which a standard ear piece is inserted. Bonded to this auditory tube is a small gauge plastic tube, through which fluid in the external auditory canal can drain and which also equilibrates pressures within the canal. The 15 cm tube increases latencies by 0.5 msec and separates stimulus artefact from the cochlear microphonics, thus enabling better visualization of these waves. The frequency characteristics of the click stimulus are not significantly altered by the tube although the stimulus output is damped by 5–7 db.

A small tympanotomy is made in the posterior inferior part of the tympanic membrane with a 21 gauge lumbar puncture needle. This small hole holds the transtympanic active electrode and tends to prevent the electrode from slipping off the convex surface of the medial promotory. The auditory tube is then placed in the external canal and both the electrode and tube are held in place with fast setting ear impression mould (Audalin). The wire is then swept forward over the face away from the operative field.

The transtympanic silver ball electrode is a low impedance electrode which provides a large response with only few sweeps so that almost constant monitoring of cochlear function can be maintained with only a 3–5-second delay. Low impedance electrodes are less affected by extraneous noise than are the higher impedance needle electrodes used by Ojemann and Levine[62].

A stimulation rate of 10 clicks/sec with a duration of 0.2 msec and a stimulus output of 90–100 db is used. The click stimulus is delivered to the auditory tube through a standard earphone (8 ohm impedance). The responses are amplified 10^4 times and 64–128 sweeps used depending on the quality of the response. Averaged waves are smoothed once, with a binomial 5-point digital filter, providing zero phase shift and an effective low-pass 3 db frequency of 1,300 Hz. The sampling interval is 10 msec for 256 points. The latency of the waves is measured from the onset of the click stimulus to each peak by digitimer 200 D computer.

At the end of the procedure, the electrode and auditory tube are easily removed with gentle traction and antibiotic eardrops instilled in the ear for the next 3 days.

Ojemann et al.[62] have described 22 consecutive patients with unilateral acoustic neuroma in whom intraoperative monitoring was performed and our own experience extends to 16 such cases. A series of stable waves forms are easily reproduced, and unlike the brain stem responses N 1 and cochlear microphonics do not appear sensitive to anesthetic drugs. Four of eight patients with good speech discrimination retained the same quality of speech function postoperatively and in the small acoustic neuromas preservation of the facial nerve of course is routine.

In common with Ojemann, we have found preservation of the N 1 and cochlear microphonics much better correlated with hearing preservation than the presence of wave V on the BAEP. It has been Ojemann's experience that all patients who could hear postoperatively had a detectable N 1 by the time the skin was closed although the potential might fade from time to time in the course of dissection of the internal auditory meatus. Indeed, in the 11 patients of Ojemann's where there was no postoperative hearing all but one lost N 1 by the end of the operation. Levine[53] has made the point that the presence of N 1 does not necessarily guarantee useful hearing. The present position appears to be that loss of N 1 and impairment of cochlear microphonics during the operation without any signs of recovery are likely to be associated with total hearing loss. However, preservation of N 1 and

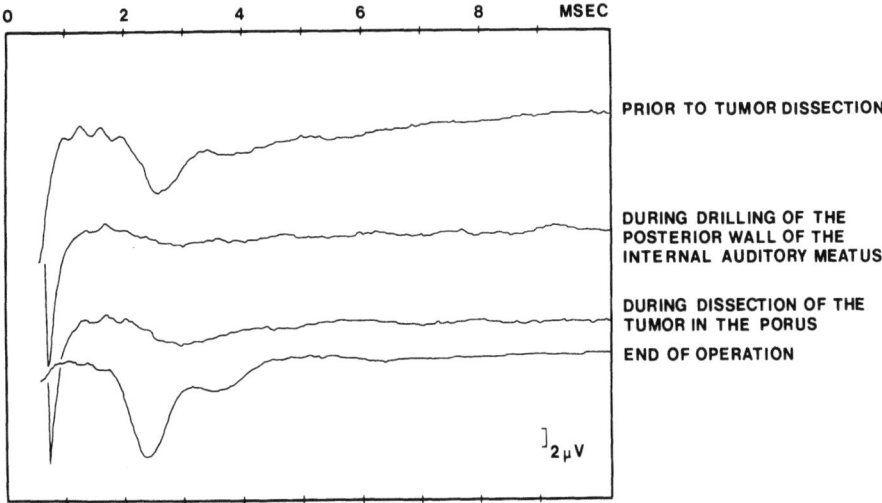

Fig. 15. Serial tracings of the electrocochleogram during acoustic neuroma surgery. Note the reduced amplitude and prolonged latency of N_1 during manipulation of the tumour

CM recordings does not necessarily guarantee good preservation of hearing.

During the dissection of the cochlear nerves, phasic changes in N_1 are not uncommon. Fig. 15 shows a good N_1 prior to tumor dissection which faded out during drilling of the posterior wall of the internal meatus and during dissection of the tumor in the porus. It had returned, however, by the end of the operation and the patient had good postoperative hearing. We have on several occasions seen the gradual disappearance of N_1 associated with coagulation of small blood vessels in the internal auditory meatus and can only conclude that under these circumstances the little vessels must have been an important part of the internal auditory artery supply. It has been our experience that coagulation of vessels within the meatus or on the nerves is best avoided, hemostasis being insured by gentle lavage and perhaps the application of thrombin or tissue glue.

Both Ojemann and Grundy[29, 62] commented that the presence of wave V of the BAEP at the conclusion of the operation generally signifies postoperative hearing, but as has been Ojemann's experience, the absence of wave V in the BAEP at the end of the operation does not by any means preclude useful hearing. Three of our patients had useful hearing despite no detectable wave V at the end of the procedure.

The present status of electrocochleography, therefore, is that this technique has a much more rapid feedback capacity because of the greater size and ease of recording of N 1 and that there appears to be a closer correlation between the N 1 and cochlear microphonic potentials and the preservation of hearing than can be obtained by any group in relation to the later potentials in the brain stem. During surgery waxing and waning of the N 1 potential may indicate potential embarrassment of some vital small vessels which under the microscope may subsequently be avoided. Of course, as Ojemann points out sudden and abrupt loss of the evoked response which does not recover is of little guide to anybody. However, this technique is likely to expand in utility as the challenge of hearing preservation for example in bilateral acoustic neuromas in neurofibromatosis is more closely addressed by more groups of neurosurgeons, and in the author's views this technique is well worth pursuing.

5. Spinal Cord Monitoring

Monitoring of evoked responses during manipulations of the spinal cord has been carried out in several circumstances in recent years.

Jones et al.[44] have reviewed their experiences in the monitoring of somatosensory evoked responses from epidural recording to posterior tibial nerve stimulation during spinal cord instrumentation in the treatment of scoliosis. They have realistically pointed out that decision about the effectiveness or otherwise of such a monitoring technique is impossible but have shown that manipulations or the cord will undoubtedly modify the evoked responses and will give warning of involvement of spinal cord blood supply in excessive distraction for example during the insertion of Harringtom rods. They observed that in comparison with data collected from a large number of cases[57] the fact that in their 410 cases, there was no incidence of postoperative paraplegia suggested that spinal cord monitoring was ethically and functionally worthwhile.

Spinal cord monitoring has also been used as an assessment of prognosis in either meningomyelocele[60] or in posttraumatic cases[76] and has proved of interest in cervical spondylosis[49] in the removal of intramedullary tumors[81], and in monitoring the progress of dorsal route entry zone lesions[46].

A full review of the present status of spinal cord monitoring is given in proceedings of two recent symposia[34] and it can be regarded as a technique of considerable potential but as yet of unproven place in the measurement of neurosurgical conditions of the cord.

References

1. Adler G, Schwerdtfeger K, Kivelitz R, Nacimiento AC, Loew F (1983) The use of somatosensory and brain-stem evoked potentials for prognostic and localization purposes in the assessment of head injury. Acta Neurochir (Wien) 68: 155–156
2. Adler G, Schwerdtfeger K, Lange E, Kivelitz R, Nacimiento AC, Loew F (1985) The use of somatosensory visual and brain-stem auditory evoked potentials for prognostic and localization purposes in the assessment of head injury. In: Rizzo PA, Morocutti C (eds) Evoked potentials—neurophyiological and clinical aspects, Elsevier, (in press)
3. Allison T (1962) Recovery function of somatosensory responses in man. Electroencephalogr Clin Neurophysiol 14: 331–343
4. Astrup J, Siesjö BK, Symon L (1981) Thresholds in cerebral ischemia—the ischemic penumbra. Stroke 12: 723–725
5. Astrup J, Symon L, Branston NM, Lassen NA (1977) Cortical evoked potential and extracellular K^+ and H^+ at critical levels of brain ischemia. Stroke 8: 51–57
6. Becker DP, Miller JD, Ward JD, Greenberg RP, Young HF, Sakalas R (1977) The outcome from severe head injury with early diagnosis and intensive management. J Neurosurg 47: 491–502
7. Bell BA, Symon L, Branston NM (1985) CBF and time threshold for the formation of ischemic cerebral edema, and effect of reperfusion in baboons. J Neurosurg 62: 31–41
8. Branston NM, Strong AJ, Symon L (1977) Extracellular potassium activity, evoked potential and tissue blood flow. J Neurol Sci 32: 305–328
9. Branston NM, Symon L (1980) Cortical SEP, blood flow and potassium change in experimental ischemia. In: Barber D (ed) Evoked potentials. Lancaster, England MTP, pp 527–530
10. Branston NM, Symon L, Crockard HA, Pásztor E (1974) Relationship between the cortical evoked potential and local cortical blood flow following acute middle cerebral artery occlusion in the baboon. Exp Neurol 45: 195–208
11. Brewer CC, Resnik DM (1984) The value of BAEP in assessment of the comatose patient. In: Nodar RH, Barber C (eds) Evoked potentials II Butterworth Publishers, Boston-London-Sydney-Wellington-Durban-Toronto, pp 578–581
12. Bruce DA, Gennarelli TA, Langfitt TW (1978) Resuscitation from coma due to head injury. Crit Care Med 6: 254–269
13. Calvet J, Cathala HP, Hirsch J, Scherer J (1956) La response de l'homme étudiée par une méthode d' integration. CR Soc Biol (Paris) 150: 1348–1351
14. Casler JA, Hoffman R, Berger L, Billinger TW, Sinus JK, Bickford RG (1973) Use of photo diode stimulations in clinical and experimental electroencephalography and electroretinography. Electroencephalogr Clin Neurophysiol 34: 437–439
15. Ciganek L (1961) The EEG response (evoked potential) to light stimulus in man. Electroencephalogr Clin Neurophysiol 13: 165–172

16. Cloward RB (1958) The anterior approach for removal of ruptured cervical discs. J Neurosurg 15: 602–617
17. Costa e Silva IE, Wang AD, Symon L (1985) The applications of flash visual evoked potentials during operations on the anterior visual pathways. Neurol Res 7: 11–16
18. Desmedt JE, Cheron G (1980) Central somatosensory conduction in man: Neural generators and interpeak latencies of the far field components recorded from the neck and right and left scalp and earlobes. Electroencephalogr Clin Neurophysiol 50: 382–403
19. Desmedt JE, Cheron G (1980) Somatosensory evoked potentials to finger stimulation in healthy octogenarians and in young adults: Wave forms, scalp topography and transit times of parietal and frontal components. Electroencephalogr Clin Neurophysiol 50: 404–425
20. Feinsod M, Selhorst JB, Hoyt WF, Wilson WB (1976) Monitoring optic nerve function during craniotomy. J Neurosurg 44: 29–31
21. Ferguson GG, Harper AM, Fitch W, Rowan JO, Jennett B (1972) Cerebral blood flow measurements after spontaneous subarachnoid hemorrhage. Eur Neurol 8: 15–22
22. Fox JE, William B, 1984: Central conduction following surgery for cerebral aneurysm. J Neurol Neurosurg Psychiatry 47: 873–875
23. Gennarelli TA, Spielman GM, Langfitt TW, Gildenberg PL, Harringtom T, Jane JA, Marshall LF, Miller JD, Pitts LH (1982) Influence of the type of intracranial lesion on outcome from severe head injury. J Neurosurg 56: 26–32
24. Goldie WD, Chiappa KH, Young RR, Brooks EB (1981) Brainstem auditory and short-latency evoked responses in brain death. Neurology (NY) 31: 248–256
25. Greenberg RP, Becker DP, Miller JD, Mayer DJ (1977) Evaluation of brain function in severe head trauma with multimodality evoked potentials. Part 2, Localization of brain dysfunction and correlation with post-traumatic neurological conditions. J Neurosurg 47: 163–177
26. Greenberg RP, Mayer DJ, Becker DP, Miller JD (1977) Evaluation of brain function in severe human head trauma with multimodality evoked potentials. Part 1, Evoked brain-injury potentials, methods, and analysis. J Neurosurg 47: 150–162
27. Grubb RL Jr, Raichle ME, Eichling JO, Gado MH (1977) Effect of subarachnoid hemorrhage on cerebral blood volume, blood flow, and oxygen utilization in humans. J Neurosurg 46: 446–453
28. Grundy BL, Nelson PB, Lina A, Heros RC (1982) Monitoring of cortical SSEP to determine safety of sacrificing the anterior cerebral artery. Neurosurgery 11: 64–67
29. Grundy BL, Yanetta PJ, Procopio PhT, Lina A, Boston JR, Dayle E (1982) Intraoperative monitoring of brainstem auditory evoked potentials. J Neurosurg 57: 674–681
30. Halliday AM, Halliday E, Kriss A, McDonald WI, Mushin J (1976) The pattern-evoked potentials in compression of anterior visual pathways. Brain 99: 357–374
31. Harding GFA (1977) The use of visual evoked potentials to flash stimuli in the

diagnosis of visual defects. In: Desmedt JE (ed) Visual evoked potentials in man: new developments. Clarendon Press, London, pp 500–508
32. Harris RJ, Symon L (1984) Extracellular pH, potassium, and calcium activities in progressive ischemia in rat cortex. Cereb Blood Flow Metab 4: 178–186
33. Heiss WD, Hayakawa T, Waltz AG (1976) Cortical neuronal function during ischemia. Arch Neurol 33: 813–820
34. Homma S, Tamaki T (eds) (1984) Fundamentals and clinical application of spinal cord monitoring. Saikon Publishing, Tokyo
35. Hume AL, Cant BR (1978) Conduction time in somatosensory pathways in man. Electroencephalogr Clin Neurophysiol 45: 361–375
36. Hume AL, Cant BR (1981) Central somatosensory conduction after head injury. Ann Neurol 10: 411–419
37. Hume AL, Cant BR, Shaw NA (1979) Central somatosensory conduction time in comatose patients. Ann Neurol 5: 379–384
38. Hume AL, Cant BR, Shaw NA, Cowan JC (1982) Central somatosensory conduction time from 10 to 79 years. Electroencephalogr Clin Neurophysiol 54: 49–54
39. Hunt WE, Hess RM (1968) Surgical risk as related to time of intervention in the repair of intracranial aneurysms. J Neurosurg 28: 14–20
40. Jacobson JH, Hirose T, Suzuke TA (1968) Simultaneous ERG and VER in lesions of the optic pathway. Invest Ophtalmol 7: 279–292
41. James IM (1968) Changes in cerebral blood flow and in systemic arterial pressure following spontaneous subarachnoid hemorrhage. Clin Sci 35: 11–22
42. Jennett B, Bond M (1975) Assessment of outcome after severe brain damage—a practical scale. Lancet 1: 480–484
43. Jennett B, Teasdale G, Galbraith S, Pickard J, Grant H, Braakman R, Avezaat C, Maas A, Minderhead J, Vecht CJ, Heiden J, Small R, Caton W, Kurze T (1977) Severe head injuries in three countries. J Neurol Neurosurg Psychiatry 40: 291–298
44. Jones SJ, Carter L, Edgar MA, Morley T, Ransford AO, Webb PJ (1985) Experience of epidural spinal cord monitoring in 410 cases. In: Schramm J, Jones SJ (eds) Spinal cord monitoring. Springer, Berlin Heidelberg New York Tokyo, pp 221–226
45. Jones TA, Stockard JJ, Henry KR (1978) Temperature-independent alterations of brainstem auditory evoked responses by enflurane. Soc Neurosci Abstr 4: 154
46. Jones SJ, Thomas DGT (1985) Assessment of long sensory tract conduction in patients undergoing dorsal root entry zone coagulation for pain relief. In: Schramm J, Jones SJ (eds), Spinal cord monitoring. Springer, Berlin Heidelberg New York Tokyo, pp 266–273
47. Kálmánchey R, Avila A, Symon L (1986) The use of brainstem auditory evoked potentials during posterior fossa surgery. Acta Neurochir (Wien) 82: 128–136
48. Kidooka M, Watanabe K, Matsuda M, Handa J (1986) Monitoring of somatosensory voked potentials during aneurysm surgery. Surg Neurol (in press)

49. Kotani H, Hattori S, Senzoku F, Kawai S, Saiki K, Yamasaki H, Omote K (1985) Evaluation of cord function in cervical spondylosis by a combined method using segmental and conductive spinal evoked potentials (SEP). In: Schramm J, Jones SJ (eds), Spinal cord monitoring. Springer, Berlin Heidelberg New York Tokyo, pp 274–283
50. Kriss A (1982) Stimulating techniques and recording problems. In: Halliday AM (ed) Evoked potentials in clinical testing. Churchill Livingstone, Bath, England, pp 45–70
51. Ladds A, Branston NM, Symon L (1984) Changes in the SEP in thalamus during a selective thalamic ischemic lesion. Electroencephalogr Clin Neurophysiol 58: 114
52. Larson SJ, Sances A Jr, Ackmann JJ, Reigel DH (1973) Non-invasive evaluation of head trauma patients. Surgery 74: 34–40
53. Levine RA, Ojemann RG, Montgomery WM, McGaffigan P (1984) Monitoring of auditory evoked potentials during acoustic neuroma surgery: Insight into mechanisms of the hearing loss. Ann Otol Rhinol Laryngol 93: 116–123
54. Lindsay KW, Carlin J, Kennedy I, Fry J, McInnes A, Teasdale GM (1981) Evoked potentials in severe head injury-analysis and relation to outcome. J Neurol Neurosurg Psychiatry 44: 796–802
55. Merory J, Thomas DJ, Humphrey PRD, Du Boulay GH, Marshall J, Ross Russell RW, Symon L, Zilkha E (1980) Cerebral blood flow after surgery for recent subarachnoid hemorrhage. J Neurol Neurosurg Psychiatry 43: 214–221
56. Meyer CHA, Lowe D, Meyer M, Richardson PL, Neil-Dwyer G (1983) Progressive change in cerebral blood flow during the first three weeks after subarachnoid hemorrhage. Neurosurgery 12: 58–76
57. McEwan GD, Bunnell WP, Sriram K (1975) Acute neurological complications in the treatment of scoliosis: a report of scoliosis research society. J Bone Joint Surg (Am) 57-A: 404–408
58. Momma F, Wang AD, Symon L (1985) Effects of temporary arterial occlusion on somatosensory evoked responses in aneurysm surgery. Surg Neurol (in press)
59. Morawetz RB, de Girolami U, Ojemann RG, Marcoux RW, Crowell RM (1978) Cerebral blood flow determined by hydrogen clearance during middle cerebral artery occlusion in unaesthetized monkeys. Stroke 9: 143–149
60. Nakagawa T, Imai K, Murakami M, Inoue SI, Maie M, Yamane T, Yamashita T (1985) Spinal evoked potentials in patients with meningomyelocele. In: Schramm J, Jones SJ (eds) Spinal cord monitoring. Springer, Berlin Heidelberg New York Tokyo, pp 231–236
61. Newlon PG, Greenberg RP, Hyatt MS, Enas GG, Becker DP (1982) The dynamics of neuronal dysfunction and recovery following severe head injury assessed with serial multimodality evoked potentials. J Neurosurg 57: 168–177
62. Ojemann RG, Levine RA, Montogomery WM, McGaffigan P (1984) Use of intraoperative autidory evoked potentials to preserve hearing in unilateral acoustic neuroma removal. J Neurosurg 61: 938–948
63. Overgaad J, Christensen S, Hvid-Hansen O, Haase J, Land AM, Hein O, Pedersen KK, Tweed WA (1973) Prognosis of head injury based on early clinical examination. Lancet 2: 631–635

64. Raudzens PA (1982) Intraoperative monitoring of evoked potentials. Ann NY Acad Sci 388: 308–326
65. Rosenstein J, Wang AD, Symon L, Suzuki M (1985) Relationship between hemispheral CBF, CCT, and clinical grade in aneurysmal subarachnoid hemorrhage. J Neurosurg 62: 25–30
66. Sachs L (1984) Angewandte Statistik. 6. Aufl. Springer, Berlin Heidelberg New York Tokyo
67. Seals DM, Rossiter VS, Weinstein ME, Spencer JD (1979) Brainstem auditory evoked responses in patients comatose as a result of blunt head trauma. J Trauma 19: 347–353
68. Starr A, Achor LJ (1975) Auditory brain stem responses in neurological disease. Arch Neurol 32: 761–768
69. Sundt TM Jr, Sharbrough FM, Anderson PE, Michenfelder JD (1974) Cerebral blood flow measurements and electroencephalograms during carotid endarterectomy. J Neurosurg 41: 310–320
70. Symon L (1985) Flow thresholds in brain ischemia and the effects of drugs. Br J Anesth 57: 34–43
71. Symon L (1985) Threshold of ischemia applied to aneurysm surgery. Acta Neurochir (Wien) 77: 1–7
72. Symon L, Ackerman R, Bull JW, Du Boulay GH, Marshall J, Rees JE, Ross Russell RW (1972) The use of xenon clearance method in subarachnoid hemorrhage. Eur Neurol 8: 8–14
73. Symon L, Brierley J (1976) Morphological changes in cerebral blood vessels in chronic ischaemic infarction flow correlation obtained by the hydrogen clearance method. In: Cervos-Navarro J, Matakas F (eds) The Cerebral Vessel Wall Symposium, Berlin, March 14–15, 1975, Raven Press, New York, pp 165–174
74. Symon L, Jakubowsky J (1979) Transcranial management of pituitary tumours with suprasellar extention. J Neurol Neurosurg Psychiatry 42: 123–133
75. Symon L, Wang AD, Silva IEC, Gentili F (1984) Perioperative use of somatosensory evoked responses in aneurysm surgery. J Neurosurg 60: 269–275
76. Takano H, Takakuwa K, Tsuji H, Nakagawa T, Imai K, Inoue S (1985) An assessment of the use of spinal cord evoked potentials in prognosis estimation of injured spinal cord. In: Schramm J, Jones SJ (eds), Spinal cord monitoring. Springer, Berlin Heidelberg New York Tokyo, pp 221–226
77. Tawana LK, Pickard JD, Sedgwick EM, Docherty TB (1983) Changes in evoked potentials and clinical tests of vision in relation to pituitary surgery. Presented during the first Joint Meeting of the EEG Society and Psychophysiology Society. Burden Neurological Institute, 29th June–1st July, 1983
78. Teasdale G, Jennett B (1976) Assessment of coma and impaired consciousness. A practical scale. Lancet 2: 81–84
79. Trojaborg W, Boysen G (1973) Relation between EEG, regional cerebral blood flow and internal carotid endarterectomy. Electroencephalogr Clin Neurophysiol 34: 61–69
80. Uhl RR, Squires KC, Bruce DL, Starr A (1980) Variation in visual evoked

potentials under anesthesia. In: Kornhuber HH, Deecke L (eds) Progress in brain research, Vol. 54. Elsevier, North Holland, pp 463–466
81. Valencak E, Witzmann A, Reisecker F (1985) Intraoperative spinal cord monitoring at differential levels and with varying surgical pathology. In: Schramm J, Jones SJ (eds), Spinal cord monitoring. Springer, Berlin Heidelberg New York Tokyo, pp 237–244
82. Walser H, Yaşargil MG, Curcic M (1982) Auditory brainstem responses in patients with posterior fossa tumours. Surg Neurol 18: 405–415
83. Wang AD, Cone J, Symon L, Silva IEC (1984) Somatosensory evoked potential monitoring during the management of aneurysmal SAH. J Neurosurg 60: 264–268
84. Wang AD, Silva IEC, Symon L, Jewkes D (1985) The effects of halothane on somatosensory evoked potentials during operations. Neurol Res 7: 58–62
85. Wilson WB, Kirsch WN, Neville H, Stears J, Feinsod M, Lehman RAW (1976) Monitoring of visual function during parasellar surgery. Surg Neurol 5: 323–329
86. Wright JE, Arden G, Jones BR (1973) Continuous monitoring of visually evoked responses during intraorbital surgery. Trans Ophthalmol Soc UK 93: 311–314
87. Schwerdtfeger K, Ludt H (1986) Stainless steel skin staples—a useful electrode system for long-term electrophysiological measurements in neurosurgery. Acta Neurochir (Wien) 82: 137–140

The Biological Role of Hypothalamic Hypophysiotropic Neuropeptides[1,2]

K. VON WERDER

Medizinische Klinik Innenstadt, University of Munich,
Munich (Federal Republic of Germany)

With 18 Figures

Contents

1. Introduction	73
2. Thyrotropin Releasing Hormone (TRH)	75
2.1. Distribution of TRH	75
2.2. TRH as a Hypophysiotropic Hormone	77
2.3. Pathophysiology of TRH	78
2.4. Clinical Utilization of TRH	78
2.4.1. Diagnosis of Thyroid Disorders	78
2.4.2. TRH in the Diagnosis of Hypothalamic-Pituitary Disease	80
2.4.3. Diagnostic Use of TRH in Disorders of PRL Secretion	82
2.4.4. TRH as a Diagnostic Aid in Acromegaly	83
2.4.5. Therapeutic Aspects of TRH	84
3. Gonadotropin Releasing Hormone (GnRH)	84
3.1. Distribution of GnRH	85
3.2. GnRH as a Hypophysiotropic Hormone	85
3.2.1. Physiological Role of Endogenous GnRH	85
3.2.2. Stimulation of Gonadotropin Secretion with GnRH	86
3.3. Pathophysiology of GnRH	88
3.4. Clinical Utilization of GnRH	89
3.4.1. GnRH in the Diagnosis of Gonadal Disorders	89
3.4.2. GnRH Test in Hyperprolactinemic Disorders	90
3.4.3. GnRH in Acromegaly	90
3.4.4. Treatment with GnRH	90
3.4.5. Superactive GnRH Agonists	92

[1] Dedicated to Prof. M. M. Forell for his seventieth birthday.
[2] Supported by the Deutsche Forschungsgemeinschaft (We 439/5–1).

4. Corticotropin Releasing Hormone (CRH) 94
　4.1. Structure of CRH 94
　4.2. Distribution of CRH 95
　4.3. Measurement of CRH 95
　4.4. CRH as a Hypophysiotropic Hormone 96
　4.5. Pathophysiology of CRH 97
　4.6. Clinical Utilization of CRH 98
　　4.6.1. Use of CRH in the Differential Diagnosis of Adrenal Failure   98
　　4.6.2. Use of CRH as a Diagnostic Aid in Cushing's Syndrome   100

5. Growth Hormone Releasing Hormone (GRH) 101
　5.1. Structure and Biological Activity of GRH 101
　5.2. Distribution of GRH 103
　5.3. Measurement of GRH 103
　5.4. Clinical Utilization of GRH 105
　　5.4.1. Biological Activity of GRH in Normal Subjects   105
　　5.4.2. Evaluation of Anterior Pituitary Function with GRH   107
　　5.4.3. Use of GRH in the Differential Diagnosis and Treatment of
　　　　　 Pituitary Dwarfism   109
　　5.4.4. GRH as a Diagnostic Aid in Acromegaly   109

6. Somatostatin (SRIF) 112
　6.1. Distribution of Somatostatin 113
　6.2. Physiological Role of Somatostatin 114
　　6.2.1. Somatostatin as a Hypophysiotropic Hormone   114
　　6.2.2. Measurement of Somatostatin   114
　6.3. Pathophysiology of Somatostatin 115
　6.4. Clinical Use of Somatostatin 115

7. Hypothalamic Hormones Regulating Prolactin Secretion 117
　7.1. Prolactin Inhibiting Factor 117
　7.2. Prolactin Releasing Factor 118
　7.3. Clinical Utilization of Hypothalamic Hormones Regulating PRL
　　　　 Secretion   119

8. Summary 119

References 122

List of Abbreviations

ACTH　　Adrenocorticotropic hormone
CRH*　　Corticotropin releasing hormone

---

* The hypothalamic hypophysiotropic neurohormones were labelled releasing factors (RF) as long as the structure of these neurohormones was not established. Since this has been achieved, all releasing factors are now called releasing hormones (RH) with the exception of the factors controlling PRL-secretion. In the manuscript DA is still labelled PIF, though the role of DA is unequivocally established and therefore could also be labelled PIH.

DA	Dopamine
DNA	Desoxy-ribonucleic acid
FSH	Follicle stimulating hormone
GAP	GnRH-associated peptide
GH	Growth hormone
GnRH	Gonadotropin releasing hormone
GRH	Growth hormone releasing hormone
hCG	Human chorionic gonadotropin
hMG	Human menopausal gonadotropin
LH	Luteinizing hormone
OT	Oxytocin
PIF*	Prolactin inhibiting factor
POMC	Proopiomelanocortin
PRF*	Prolactin releasing factor
PRL	Prolactin
RNA	Ribonucleic acid
SRIF	Somatotropin release inhibiting factor
SS 1-14	Somatostatin 1-14
TRH	Thyrotropin releasing hormone
TSH	Thyroid stimulating hormone
VIP	Vasoactive intestinal polypeptide

1. Introduction

Four decades ago, Sir Geoffrey Harris postulated that the anterior pituitary is under hypothalamic neurohormonal control[52] thus linking the brain with the endocrine system. This concept has led to a very fruitful search for the neurohormonal substances of the hypothalamus, which reach the anterior pituitary via the portal system, a vascular short cut between the neural tissue of the brain and the endocrine tissue of the pituitary[113] (hypophysiotropic peptides), or which are directly secreted into the general circulation (posterior lobe peptides).

These peptides are called neurohormones since they are secreted from neuroendocrine cells which combine the typical features of neuronal cells—electric excitation and synaptic connections within the central nervous system—with cells of endocrine glands—secretion of a substance into the blood stream. Though the neuropeptides have been initially discovered as products of such neuroendocrine cells, it is clear now that these substances are also produced by purely neuronal cells in the CNS where they may have behavioral and psychotropic effects[193]. Furthermore, they are produced in the peripheral nervous system and also by purely endocrine cells located in peripheral endocrine glands. In addition, peptides initially identified as hormones secreted from peripheral endocrine glands were found to be produced by neuronal cells in the CNS and secreted into the synaptic cleft, thus abolishing the previous strict separation between endocrine and neuronal transmission of messages[48].

Whereas the structure of the two posterior lobe nonapeptides had already been elucidated, their synthesis performed and their genes cloned in the late fifties[130], the structure of the first hypophysiotropic peptides, thyrotropin releasing hormone (TRH) and gonadotropin releasing hormone (GnRH), a tri- and a decapeptide respectively, were elucidated only in the early seventies [18, 19, 92]. While it was clear that there must be a

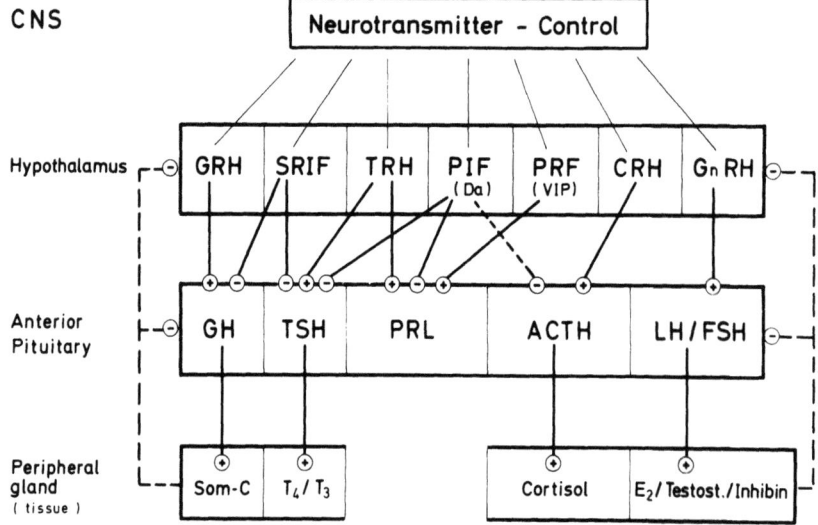

Fig. 1. Control of anterior pituitary function by the hypothalamic hypophysiotropic neurohormones

neurohormone stimulating the release of growth hormone and ACTH, they remained elusive. While looking for growth hormone releasing factor activity, an inhibiting hormone for growth hormone release, somatostatin, a cyclic peptide consisting of fourteen amino acids, was discovered[15]. Only 4 and 5 years ago, respectively, were the long chain neurohormones corticotropin releasing hormone (CRH) and growth hormone releasing hormone (GRH) discovered, their structure elucidated and their synthesis successfully completed[49, 173, 174]. Though there is still uncertainty about the neuropeptides regulating prolactin secretion, it has become clear that the main factor in the regulation of this hormone is dopamine, which is also secreted into the portal circulation, exerting an inhibitory effect on prolactin secretion[6]. Today, with the the exception of the neurohormones for prolactin regulation, the structure of the neuropeptides regulating the secretion of the anterior pituitary hormones is known; their synthesis has been completed and in most instances their respective genes coding the neuropeptide precursors have been cloned[39, 45, 47, 151]. Potent analogues and antagonists have been synthesized by aminoacid substitution or deletion and these analogues as well as the native peptides have become important agents in diagnosing and treating endocrine as well as non endocrine diseases[139, 180]. Thus, the TRH-test has become an important tool in the diagnosis of thyroid disorders and the CRH test may prove to be of value for the differential diagnosis of hyperadrenocorticism. Furthermore, it is hoped that the administration of hypothalamic peptides might help to differentiate between hypothalamic and pituitary causes of anterior pituitary failure. This has turned out to be more difficult than it was initially thought on a theoretical basis. Somatostatin has never achieved any diagnostic importance though it has become an important therapeutic agent. The therapeutic indications for GnRH and probably GRH are established whereas CRH has up to now no therapeutic value, and therapy with TRH remains in the experimental stage.

In the following review the biological significance of the hypothalamic neuropeptides involved in regulation of anterior pituitary function (Fig. 1) will be discussed with particular emphasis upon their diagnostic and therapeutic role in clinical medicine.

2. Thyrotropin Releasing Hormone (TRH)

Thyrotropin releasing hormone (TRH) was purified from ovine hypothalamic extracts based on its ability to induce TSH-secretion *in vivo*[19]. The structure was an unusual tripeptide with blocked N- and C-terminal residues, L-pyroglutamil-L-histidyl-L-proline amide (Table 1), a structure which was later identified also in man [17, 58, 60]. Since the short tripeptide could be synthesized easily, the early availability of the synthetic peptide led to a rapid accumulation of information in respect to distribution, biological function and clinical application of this releasing hormone.

2.1. Distribution of TRH

The highest concentration of TRH has been found in the hypothalamus, particularly in the median eminence region[16, 61]. However, significant concentrations of TRH are also found in extrahypothalamic sites of the CNS. Though the concentration of TRH in the extrahypothalamic rat brain is low compared to levels found in the hypothalamus, more than 70% of total rat brain TRH is found outside the hypothalamic region[61]. Furthermore, TRH is found in extraneuronal tissue, particularly in the gastrointestinal tract and in the pancreas, where it has been localized in the islets of Langerhans[46]. Immunoreactive TRH-like material has also been reported in the reproductive system of the male rat, including the prostate[46].

Though the hypophysiotropic role of hypothalamic TRH is well

Table 1. *Hypothalamic Hormones and Peptides*

	Structure	Function
1. Hypophysiotropic hormones		
TRH Thyrotropin-releasing hormone	pyro-Glu-His-Pro-NH$_2$	stimulates TSH- and PRL-secretion
GnRH Gonadotropin-releasing hormone	pyro-Glu-His-Trp-Ser-Tyr-Gly-Leu-Arg-Pro-Gly-NH$_2$	stimulates LH- and FSH-secretion
CRH Corticotropin-releasing hormone	H-Ser-Glu-Glu-Pro-Pro-Ile-Ser-Leu-Asp-Leu-Thr-Phe-His-Leu-Leu-Arg-Glu-Val-Leu-Glu-Met-Ala-Arg-Ala-Glu-Gln-Leu-Ala-Gln-Gln-Ala-His-Ser-Asn-Arg-Lys-Leu-Met-Glu-Ile-NH$_2$	stimulates ACTH-secretion
GRH Growth hormone releasing hormone	H-Tyr-Ala-Asp-Ala-Ile-Phe-Tyr-Asa-Ser-Tyr-Arg-Lys-Val-Leu-Gly-Gln-Leu-Ser-Ala-Arg-Lys-Leu-Leu-Gln-Asp-Ile-Met-Ser-Arg-Gln-Gln-Gly-Glu-Ser-Asn-Gln-Glu-Arg-Gly-Ala-Arg-Ala-Arg-Leu-NH$_2$	stimulates hGH-secretion
SRIF STH(GH)-release inhibiting hormone, somatostatin	H-Ala-Gly-Cys-Lys-Asn-Phe-Phe-Trp-Lys-Thr-Phe-Thr-Ser-Cys-OH-S-S	inhibits hGH- and TSH-secretion
PIF Prolactin inhibiting factor	dopamine	Inhibits prolactin- and TSH-secretion (in acromegalics also hGH)
PRF Prolactin releasing factor (= VIP, vasoactive intestinal peptide)	His-Ser-Asp-Ala-Val-Phe-Thr-Asp-Asn-Tyr-Thr-Arg-Leu-Arg-Lys-Gln-Met-Ala-Val-Lys-Lys-Tyr-Leu-Asn-Ser-Ile-Leu-Asn-NH$_2$	stimulates prolactin-secretion
2. Posterior lobe peptides		
ADH Antidiuretic hormone	Cys-Tyr-Phe-Gln-Asn-Cys-Pro-Arg-Gly-NH$_2$	stimulates water resorption in the kidney
OT Oxytocin	Cys-Tyr-Ile-Gln-Asn-Cys-Pro-Leu-Gly-NH$_2$	stimulates uterine contraction and milk ejection

known[99], the function of TRH in the CNS as a neurotransmitter is still unclear. Neurophysiologic studies show that TRH effects electrical activity of neurons and influences the excitatory or depressive effects of other neurotransmitters. Still more elusive is the role of TRH in the gastrointestinal tract, in the pancreas, and in the gonadal system. As all hypophysiotropic hormones, TRH has also been found in the placenta, again the functional role being unknown[46, 157].

2.2. TRH as a Hypophysiotropic Hormone

Biosynthesis of TRH in the hypothalamus involves, as with all other hypothalamic peptides, a high molecular weight precursor, which is produced in the endoplasmic reticulum and later transferred to its site of release[46]. The exact biosynthetic pathway has not been completely elucidated and in contrast to the other hypophysiotropic hypothalamic hormones, synthesis and cloning of the DNA complementary to the messenger RNA for TRH prohormone has not been achieved[46].

TRH is released into the pituitary portal system and binds to TRH receptors located at the membrane of the thyrotrophs[53]. The receptors are linked to the adenyl cyclase system. Thus, exposure of thyrotrophs to TRH leads to accumulation of cyclic AMP and later to stimulation of TSH secretion. However, the main action of TRH is not mediated by cyclic AMP, but through break down of phosphatidyl inositolphosphates and changes of calcium fluxes[46]. The decrease of membrane bound calcium is accompanied by an increase of calcium ions in the cytoplasm. The latter seem to exert their effect, *i.e.*, synthesis and secretion of TSH and PRL with the participation of calmodulin[46]. As has been shown for other neuropeptides, TRH is also not completely specific in respect to its hypophysiotropic action. Thus, it stimulates prolactin release *in vivo* and *in vitro*. Accordingly, TRH receptors have been demonstrated at the pituitary lactotrophs[53]. Often, a considerable degree of concordance between TSH and PRL-secretion can be observed. Thus, estrogen treatment leads to higher TSH and PRL responses after TRH-administration[58, 62] and in patients with hypothyroidism, TSH as well as PRL responses to TRH are exaggerated[36, 159]. However, it is doubtful that TRH is the physiological prolactin releasing factor, since episodes of prolactin secretion which are most likely included by a hypothalamic stimulus (for example suckling induced PRL-secretion), are not accompanied by TSH-increases[36, 128]. Recently, evidence has accumulated, that vasoactive intestinal polypeptide (VIP) may be the physiologic prolactin releasing factor[1].

In addition, TRH can lead to an increase of growth hormone in pathological situations (for example acromegaly, depressive illness, breast cancer, and renal failure[32, 84, 96, 119]).

2.3. Pathophysiology of TRH

The most important effect of TRH in humans is stimulating TSH secretion. Thus, experimental and pathological disruption of the hypothalamic blood supply to the pituitary results in decreased thyroid function—hypothalamic, tertiary hypothyroidism. The latter has been observed as a sequel of hypothalamic tumours like craniopharyngiomas, dermoid cysts or primary pituitary tumours with suprasellar extension[119] but it is rarely due to an idiopathic neuronal TRH deficit[125]. While it is quite clear that lack of TRH activity, for example induced by TRH-antibodies in experimental animals, leads to thyroid failure, there are no studies reporting on actual decrease of TRH levels in the portal circulation in patients with tertiary hypothyroidism.

Since TSH-levels are suppressed in patients with hyperthyroidism, it is clear that TRH and TSH are not involved in the overwhelming majority of patients with elevated thyroid hormone levels. Very rarely TSH-induced hyperthyroidism due to a thyrotroph tumour occurs. In the present state it cannot be decided with certainty if these tumors are autonomous or of hyperplastic origin due to constant hypothalamic stimulation. This can only be assumed in hypothyroid patients with a large sella turcica and thyrotroph hyperplasia, occasionally extending into the suprasellar area and causing visual field defects, due to lack of feed back inhibition because of primary thyroid failure. However, in most patients with endocrinologically active pituitary adenomas, the differential diagnosis between hypothalamic dependence or pituitary autonomy can not be made, because the measurement of releasing hormones in the peripheral circulation does not reflect hypothalamic neurohormone secretion. This pertains particularly to TRH, which has a very short half life and cannot be measured in the peripheral circulation at all[46]. Releasing hormones can often be measured in the peripheral blood without difficulty if produced ectopically. However, ectopic TRH production has not yet been described[46].

2.4. Clinical Utilization of TRH

2.4.1. Diagnosis of Thyroid Disorders

When 200–400 µg TRH are given *i.v.*, to normal subjects, TSH increases from low or undetectable to peak levels within 15 to 30 minutes[161]. The TRH-induced TSH-response leads to an increase in triiodothyronine within 2 hours whereas the less pronounced thyroxine increase occurs later.

Intravenous injection of TRH causes transient elevation of blood pressure[14] and symptoms including nausea, flushing, or a desire to urinate in about half of the subjects: these symptoms may last up to one or two minutes. Very rarely, severe side effects with loss of consciousness and convulsions have been observed[22, 27]. Repeated injections of TRH have no effect on bone marrow, renal and hepatic function, showing that TRH has practically no toxicity[17].

There is a dose-response relationship between the TRH dosage and TSH peak response levels. The minimal TRH-dosage leading to a TSH response is 15 µg, and the maximal response is achieved with 400 µg *i.v.* in normal subjects. Higher dosages do not lead to a further increase of TSH secretion[161]. Continuous infusion of TRH leads to a biphasic TSH-increase, which cannot be further increased if TRH pulses are administered during the TRH-infusion[17, 46].

Female euthyroid subjects tend to have higher TSH responses to TRH than men. Furthermore, older men, but not women, show a decrease of the TRH-induced TSH response with increasing age[17].

TRH can also be given intranasally or orally, though 100 times higher dosages are necessary to reproduce the effect of the *i.v.* route.

Though there are numerous extrathyroidal factors influencing TSH secretion, the most important are circulating free thyroid hormones. Since basal TSH levels are already very low in euthyroid subjects, one cannot distinguish normal basal from suppressed TSH secretion using commonly available radioimmunoassays. However, suppressed and normal TSH secretion can be differentiated using standard assays if TRH-stimulated TSH-values are evaluated. Thus, in thyrotoxicosis, the TSH response to TRH is suppressed (Fig. 3) and this test therefore may serve as confirmation of the diagnosis, particularly in borderline cases[51, 158]. As has been mentioned, very rarely thyrotroph pituitary adenomas lead to thyrotoxicosis with elevated TSH levels which may or may not rise further after *i.v.* administration of TRH[56, 66]. Generally, a normal rise in TSH secretion after TRH rules out the diagnosis of hyperthyroidism. However, suppressed TSH secretion does not always indicate thyrotoxicosis or even a tendency to develop hyperthyroidism. In addition to persistent suppression of TSH secretion observed after reaching low basal thyroid hormone levels in patients with Graves' disease treated with antithyroid drugs, it is now well recognized that nonthyroidal disorders like depression[44], renal failure[184] and starvation will also lead to blunted, though rarely completely suppressed TSH responses to TRH[119]. Furthermore, the nonthyroidal endocrine status can influence the TRH test. Thus, patients with endogenous or exogenous hypercorticism may have blunted TRH tests[28, 185], and in patients with elevated growth hormone levels the TSH response to TRH may be impaired[135], though complete suppression suggests the presence of additional thyroid disease. In patients with primary thyroid failure a TSH hyperresponse to TRH is characteristically observed, though basal levels of TSH are usually already elevated (Fig. 3). However, in 20% of the patients with endemic goiters in areas of nutritional iodine deficiency who have normal basal thyroid hormone levels and normal basal TSH levels, a hyperresponse of TSH is observed, indicating a very early form of insufficiency in thyroid function[119]. Therefore, comparable to the suppressed TSH secretion in hyperthyroidism, the exaggerated TSH response to TRH in primary hypothyroidism serves as an amplifier for detecting minute decreases in circulating thyroid hormone bioactivity. Recently, supersensitive TSH radiometric and enzyme-immunoassays using monoclonal antibodies have been introduced, allowing differentiation of suppressed basal TSH levels in hyperthyroid patients from normal basal TSH levels in euthyroid subjects[10], thus making the TRH test superfluous in patients with hyperthyroidism.

2.4.2. TRH in the Diagnosis of Hypothalamic-Pituitary Disease

Theoretically, administration of TRH should allow one to make the differential diagnosis between hypothalamic and pituitary origin of secondary hypothyroidism. One would expect that patients with hypothalamic hypothyroidism would have a normal response of TSH to TRH whereas patients with low peripheral thyroid hormone levels due to pituitary tumours should have no TSH rise after TRH administration[186].

In fact, patients with hypothalamic hypothyroidism have the expected TSH response to TRH which is often delayed and sometimes exaggerated[5, 33, 119]. However, hypothyroid patients with pituitary tumours, suprasellar extension, and visual field defects (and only patients with such big tumours have secondary hypothyroidism) also show a TSH-response to TRH comparable to the one which is seen in patients with a radiologically documented hypothalamic lesion (Fig. 2). The failure of the TRH test to differentiate between hypothalamic and pituitary lesions leading to secondary hypothyroidism can be explained by the fact that patients with primary pituitary tumours may also have a hypothalamic defect. The latter is caused by compression of the portal system by suprasellar extension of the tumour which does not allow endogenous TRH to reach the residual pituitary thyrotrophs[119]. This is also suggested by the observation that TRH given intraventricularly during neurosurgery does not cause a rise in TSH, whereas TRH given systemically leads to an increase of TSH in patients with suprasellar extending tumours[34]. The observation by Faglia and coworkers[33], who showed that triiodothyronine levels did not increase appropriately after TRH-induced elevation of TSH levels in these patients, suggests a diminished biological activity of the circulating TSH, which was later documented by a cytochemical bioassay. The same group observed an elevated β-TSH over α-TSH ratio in secondary hypothyroidism which was normalized after oral TRH treatment. These authors have recently shown that longterm treatment with oral TRH can lead to an increase of the TSH

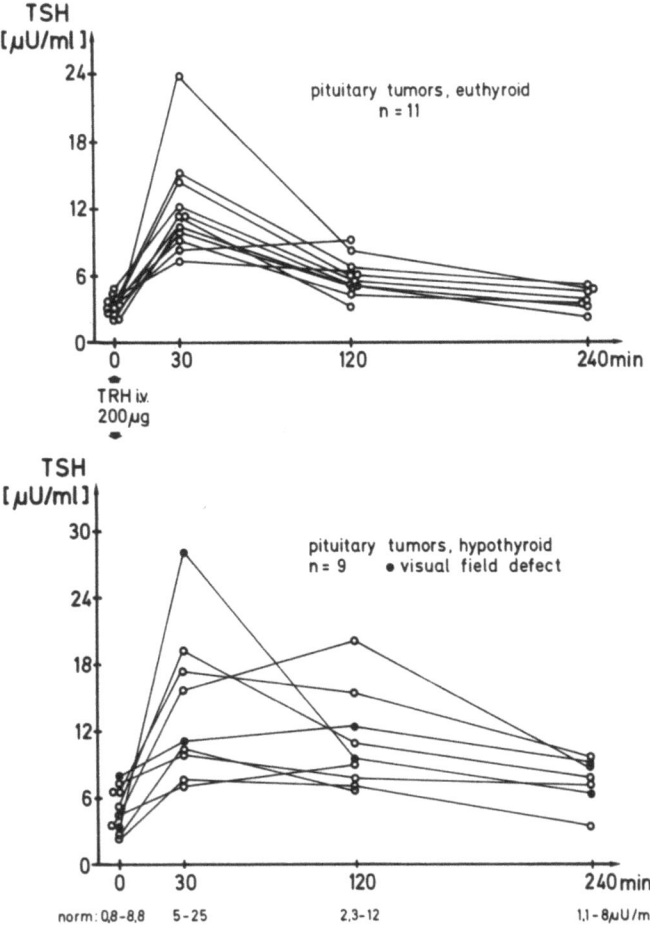

Fig. 2. TSH-response to TRH in euthyroid and hypothyroid patients with pituitary tumours. In the upper panel the normal TSH-response to TRH in euthyroid patients is shown whereas in the lower panel the more pronounced but delayed TSH-response in 9 hypothyroid patients is shown. (With permission from Pickardt and Scriba, 1985)

bioactivity and normalization of the ratio of receptor binding-activity and immunological activity of the TSH-molecule in patients with hypothalamic hypothyroidism[5]. This indicates that endogenous TRH is also necessary for assembling the correct TSH molecule with full biological activity.

Similar to the TRH-induced TSH secretion, TRH-induced PRL secretion in general does not help to differentiate between hypothalamic and pituitary lesions. Anterior pituitary failure in the presence of a radiologically normal sella turcica may be the exception. In cases with hypothalamic lesions leading to complete anterior pituitary failure, TRH can stimulate

PRL secretion, though the basal PRL level may already be elevated. In cases of pituitary autoimmune disease or apoplexy, basal levels are normal and the TRH-induced PRL increase is blunted[158].

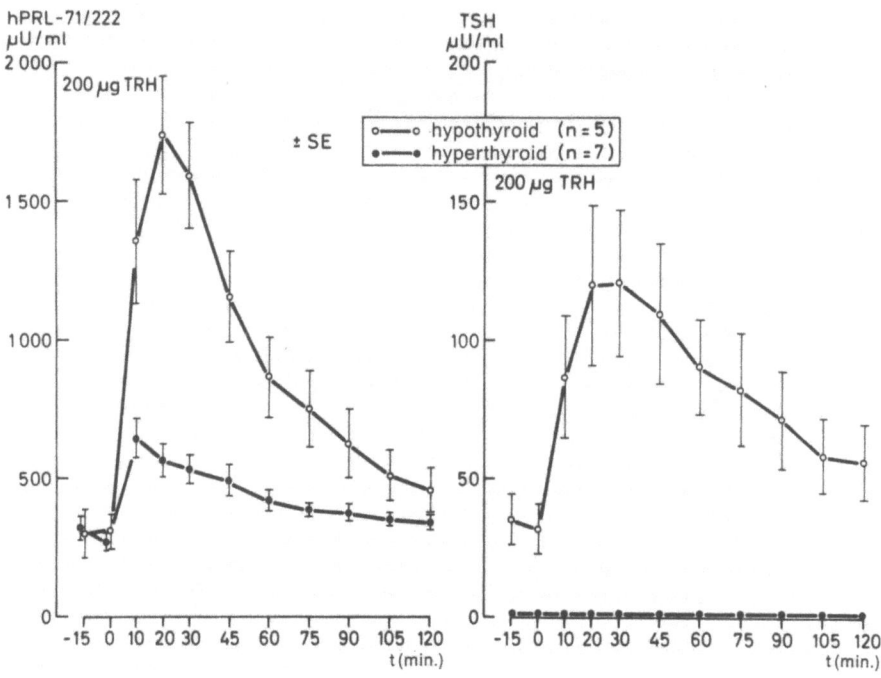

Fig. 3. Prolactin- and TSH-response to TRH in hypothyroid and hyperthyroid patients. Whereas the basal TSH-level in hypothyroidism is elevated (right panel) the basal PRL-level is within the normal range (left panel). However, there is a brisk rise of PRL after TRH above the normal range, correlating to the hyperresponse of TSH in the hypothyroid state. Typically TSH-levels are suppressed at basal state and after TRH-stimulation in hyperthyroid patients whereas there is no difference between basal PRL-levels in hypo- and hyperthyroid patients. In contrast to the complete suppression of TSH-secretion, there is a significant increase of PRL after TRH though significantly different to the PRL-rise observed in hypothyroid patients (von Werder, 1975)

2.4.3. Diagnostic Use of TRH in Disorders of PRL Secretion

As mentioned above TRH also stimulates PRL secretion. The peak rise in PRL occurs 15 to 30 min after *i.v.* administration of TRH and parallels that of TSH. Again, a dose-response relationship exists between injected TRH and stimulated PRL secretion, though in contrast to TSH, the maximal PRL increase is reached at a lower dosage (100 μg) of TRH[62]. Women who tend to have higher basal PRL levels have significantly higher

TRH-stimulated PRL levels as compared to males[62]. In patients with suppressed TRH-induced TSH secretion due to thyrotoxicosis, PRL-levels can still be stimulated (Fig. 3). In patients with primary hypothyroidism and elevated basal and TRH-induced TSH levels, basal PRL levels may still be within the normal range though a hyperresponse to TRH is observed (Fig. 3). Occasionally, patients with primary hypothyroidism have hyperprolactinemia which explains their amenorrhea and galactorrhea[36] and which responds to thyroid hormone replacement therapy[36, 159].

Patients with prolactinomas have often blunted PRL responses to TRH, though this finding is of little diagnostic value[36]. For example, some patients with macroprolactinomas have a prompt PRL rise to TRH, whereas some patients with hyperprolactinemia secondary to other causes may have no PRL response. More frequently, patients with hyperprolactinemia and a normal sella turcica have an increase of PRL after TRH which is missing in patients with radiologically documented prolactinomas[36].

This can be explained by the fact that there is an inverse correlation between the TRH-induced response and the height of the basal PRL level; the latter, in turn, roughly correlates with the size of the prolactinoma[36, 187].

2.4.4. TRH as a Diagnostic Aid in Acromegaly

TRH does not induce GH secretion after *i.v.*, intranasal, or oral administration in normal healthy subjects. However, very pronounced GH rises are observed in more than 50% of patients with active acromegaly after *i.v.* administration[32, 188]. Because of the high frequency of this abnormality, the evaluation of TRH stimulated GH secretion in acromegaly has become a diagnostic tool in addition to the oral glucose tolerance test. Furthermore, the outcome of the test seems to be of therapeutic relevance since acromegalics who have a GH increase after TRH administration usually respond to medical treatment with dopamine agonists. The fact that TRH-induced GH hyperresponses are also seen in those patients who have GH hypersecretion due to ectopic GRH production suggests that primary alteration of the receptors of the adenomatous cell does not represent a probable explanation for this abnormal response[188]. Since TRH stimulates GH secretion in other diseases such as diabetes mellitus, depression, anorexia nervosa, renal and liver failure, an altered neurotransmitter milieu of the CNS structures implicated in the control of GH secretion has to be considered[46]. Another explanation for the inappropriate GH-response to TRH in acromegaly may be that these patients have adenomas composed of somatomammotrophic cells and therefore show a "prolactin like" regulation of GH-secretion. In contrast to acromegalics, TRH does not atypically stimulate ACTH secretion from corticotrophic adenomas in patients with Cushing's disease, though this has been observed in few patients[68, 69].

2.4.5. Therapeutic Aspects of TRH

TRH is not useful in the treatment of thyroid disease. However, since in addition to its hypophysiotropic role, a general neurotransmitter role of TRH can be assumed, the administration of TRH leads to a variety of neurochemical and behavioral changes independent of the activation of the TSH-thyroid axis. Thus, TRH was found to function as an endogenous ergotropic substance[98]. It was proposed to be useful for treating psychiatric disorders[30], hastening the recovery of consciousness and respiration after anesthesia, treating narcotic overdosages, narcolepsy, brain damage and various vascular disorders[58]. The antidepressant effect, which has been reported in earlier studies has not been confirmed[63]. The reason for using TRH therapeutically in patients with spasticity was the observation that TRH is present in nerve endings in the ventral horn of lower motor neurons and in the motoric nuclei of the cerebral nerves V, VII and XII, and the fact that, in cats, a more rapid neurological recovery after spinal trauma occurred after treatment with TRH[31]. In agreement with these observations very high doses of TRH given *i.v.* to patients with amyotrophic lateral sclerosis produced marked improvement of lower and upper motor neuron deficiency with an improvement of weakness and spasticity. However, improvement was only sustained throughout the TRH infusion until one hour thereafter[29]. Furthermore, all therapeutic trials so far have been uncontrolled and the effect of TRH in human neurological disease is still subject to controversy. Another aspect to be considered is that TRH given chronically does not lead to down regulation of TSH secretion comparable to the effect of GnRH on gonadotropin release[58]. Thus, TRH induced hyperthyroidism can occur in patients given high doses of TRH chronically[119].

3. Gonadotropin Releasing Hormone (GnRH)

Gonadotropin-releasing activity was first discovered by measuring LH bioactivity released after injection of acid extracts from stalk-median eminence fragments[93]. The purification of this bioactivity led to the elucidation of the structure of the decapeptide by Schally and his group (Table 1), which paved the way for its synthesis in large quantities in the early seventies[92], which allowed a rapidly expanding number of physiological and clinical studies, leading to our present understanding of the biological and clinical, *i.e.* diagnostic and therapeutic role of GnRH.

Furthermore, potent analogues and antagonists have been developed, and some are already in clinical use[139, 180]. The complementary DNA for the messenger RNA coding the GnRH precursor peptide has recently been elucidated[107]. It is of interest, that placenta-prepro GnRH has a C-terminal peptide (GnRH-associated peptide = GAP) which contains another peptide sequence which also stimulates LH and FSH and a peptide which inhibits prolactin release. Whether the latter represents the longsought peptidergic PIF is not clear.

3.1. Distribution of GnRH

GnRH cell bodies are localized in the septal-preoptic and anterior hypothalamic area from where the axons project into the external layer of the anterior median eminence, allowing GnRH to be released into the hypophyseal portal capillaries[118, 138]. GnRH-immunoreactive neurons also project into the organum vasculosum of the lamina terminalis[93, 118] and have been found in the brainstem region, an area believed to be important for the regulation of mating behaviour[93, 101].

In addition, GnRH has been detected in other parts of the brain as well as outside the central nervous system. Thus, GnRH has been detected in the ovary and in the testes, where also receptors for GnRH could be demonstrated[93]. Furthermore, GnRH, like all other releasing hormones, is present in the placenta where it is produced as demonstrated by the presence of the GnRH messenger RNA[107].

3.2. GnRH as a Hypophysiotropic Hormone

3.2.1. Physiological Role of Endogenous GnRH

The initial event of GnRH action leading to gonadotropin release is GnRH binding to specific receptors on the surface of the gonadotrophs. The number of these GnRH binding sites seems to fluctuate under various conditions, explaining the varying responsivity of the gonadotroph to GnRH in different physiological settings (see below).

GnRH, in contrast to CRH and GRH, obviously does not act via the activation of the adenyl cyclase system[106, 160]. Thus, an increase of cAMP has not been demonstrated in *in vitro* experiments whereas cGMP was found to be increased in pituitary gonadotrophs being exposed to GnRH *in vitro*[106]. Though the guanylate cyclase system seems to play a role in mediating the biological effect of GnRH, calcium ions also seem to play an important role. Thus, the removal of calcium from the medium or addition of chelating agents inhibit GnRH-stimulated LH-release[95].

In humans with normal gonadal function, endogenous GnRH is released in a pulsatile fashion, reflected by the pulsatile secretion of LH, which can be detected by measurement of LH-levels in the peripheral blood[64, 65]. The FSH-pulses are less distinct due to the longer half-life of this hormone. The pulsatility of gonadotropin secretion is essential for normal follicular maturation, ovulation, and subsequent luteal function as well as normal testicular function[125]. Thus, in prepubertal children no pulsatility of LH can be detected, whereas at the early stage of puberty only nightly pulses of LH are observed, indicating the beginning of the activation of the central GnRH pulse generator[54, 125].

The GnRH pulses occur at 90 min intervals in females during the follicular phase, with an increasing frequency in the periovulatory period, and a slow frequency during the luteal phase. In man, between 7 and 14 pulses of LH are observed over 24 hours[179]. Neurotransmitters involved in the control of GnRH-release are dopamine and endogenous opioid peptides, both leading to a suppression of GnRH pulse amplitudes and pulse frequency[134]. The latter is reflected in the peripheral circulation by lack of LH-pulsatility[179]. Thus, heroin addicts suffer from amenorrhea and hypogonadism with irregular or low LH-pulses, which can be normalised by the administration of naloxone.

There is evidence from animal experiments, that other neurotransmitters like norepinephrine and acetylcholine are important for GnRH release, norepinephrine blocking, acetylcholine stimulating the release of GnRH. Other candidates for putative synaptic neurotransmitters regulating GnRH release are serotonin and GABA, though their precise influence in humans on gonadotropin secretion and gonadal function is not clear[93, 183].

3.2.2. Stimulation of Gonadotropin Secretion with GnRH

The injection of GnRH leads to a rapid release of LH and, to a lesser extent, of FSH. A dose-response relationship between injected GnRH and resulting gonadotropin secretion is observed from 10 µp to 3000 µg GnRH *i.v.*[11, 38].

The response of the gonadotrophs in animals and also in the human is modified by the developmental and hormonal state. Thus, injection of GnRH into prepubertal children leads to a minimal rise of LH whereas there is a normal adult type FSH increase (Fig. 4). During the peripubertal period the rise of LH after GnRH is increasing and has further increased when adulthood is reached[37]. Since the FSH-response is not significantly changing, a striking reversal of the FSH/LH ratio after GnRH-administration occurs during puberty (Fig. 4).

Furthermore, there is a sex difference in the FSH-response, *i.e.* prepubertal and pubertal females have a more pronounced rise of FSH after GnRH at all stages of sexual maturation[37]. The latter indicates that the sensitivity of the gonadotrophs to GnRH does not only change during the rapid increase of sex hormones during puberty but is also modified by the composition of the sex hormones to which the pituitary is exposed. Thus, in the adult female during the menstrual cycle, the GnRH induced gonadotropin rise is different in the early and late follicular phase and in the

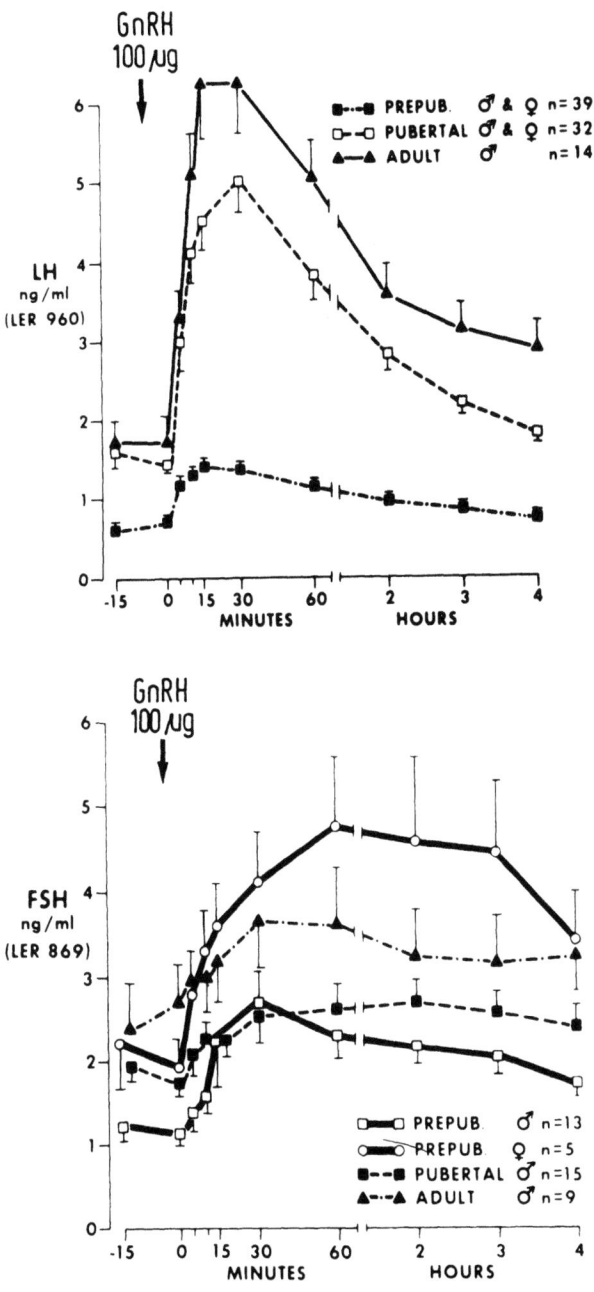

Fig. 4. LH- and FSH-responses to GnRH in different stages of sexual development. Whereas in the male no significant changes of the FSH-response to GnRH are observed, a significant increase of the maximal LH-level after GnRH stimulation occurs near puberty and after sexual maturation. This leads to a change of the FSH/LH-ratio, which is even more pronounced in females, since prepubertal girls have an even higher FSH-response before compared to after menarche. (After Grumbach *et al.*, in: Control of onset of puberty, John Wiley and Sons, New York, 1974, pp 115–166)

luteal phase, with the highest LH- and FSH-increases in the periovulatory phase[35].

The observation that continuous infusion of GnRH does not sustain elevated LH- and FSH-levels but, in contrast, leads to a decrease of the gonadotropin levels[143] has led to the concept of desensitization of the GnRH receptors at the pituitary gonadotroph level[93, 108]. The latter is the basis for the anti-gonadotropic action of GnRH superagonists, which are presently used for suppressing testicular and ovarian function[108, 139, 180]. In contrast to true GnRH-antagonists which inhibit gonadotropin secretion by competing with the native GnRH molecule for the binding sites at the GnRH receptor, the GnRH superactive analogues initially stimulate gonadotropin secretion and lead to inhibition of LH and FSH release only if treatment with GnRH super-active analogues is continued[139].

3.3. Pathophysiology of GnRH

Hypothalamic hypogonadism is caused by permanent GnRH deficiency. The latter is difficult to differentiate from delayed puberty in the age range below 20, since the hormonal abnormality in both instances is identical, *i.e.* lack of GnRH and therefore LH-pulsatility[37, 109]. In both instances baseline concentrations of LH and FSH are usually low but not infrequently may be in the normal range yet inappropriate for patients with estrogen or androgen deficiency. Only in Kallmann's syndrome, when hyposmia or anosmia is associated with GnRH-deficiency, the diagnosis of persistent hypothalamic hypogonadism can be made. The syndrome is occasionally familiar[109].

Hypogonadotropic hypogonadism, which may also present itself as isolated FSH- or isolated LH-deficiency (fertile-eunuch-syndrome) may also be due to a defect of the pituitary gonadotrophs. For differential diagnosis see page 89.

There is no clinical equivalent for excessive GnRH secretion. However, premature activation of the GnRH-pulse generator may occur leading to central true precocious puberty[109]. The latter can be idiopathic without any demonstrable anatomical lesion or symptomatic. Thus, hamartomas or pinealomas may lead to premature sexual maturation, by mechanisms which are not completely elucidated. Patients with pinealoma may suffer from precocious but also delayed puberty[109, 125].

Central true precocious puberty has to be differentiated from familiar gonadotropin independent precocious puberty and also from precocious pseudopuberty. True precocious puberty is associated with premature but otherwise completely normal GnRH and LH pulsatility whereas in the latter two situations, sexual steroids are elevated but LH is suppressed and shows no evidence of pulsatility[90].

In analogy to precocious pseudopuberty due to steroid producing tumours or congenital adrenal hyperplasia, the application of exogenous sex steroids also leads to suppression of gonadotropin levels. Whereas in the female the main effect of sex steroid feed-back takes place at the gonadotroph level, in the male androgens influence predominantly GnRH secretion, particularly suppressing pulse frequency[93].

There is no evidence for ectopic GnRH production leading to disturbance of gonadal function.

3.4. Clinical Utilization of GnRH

The GnRH test for clinical purposes is usually performed with 25, 50 or 100 µg *i.v.*, which is well tolerated and devoid of side effects[11, 186]. The maximal LH response occurs usually 15 to 30 minutes and the maximal FSH response 30 to 45 minutes after GnRH injection. Gonadal steroids, particularly estrogens and androgens have a modulatory influence on the gonadotropin response to GnRH (see above). Generally, the peak LH- and FSH-levels after GnRH injection correlate to the height of the basal level. Thus, during the prepubertal period when gonadotropin levels are low the gonadotropin response to GnRH is decreased whereas after the menopause exaggerated gonadotropin responses are observed. Furthermore, whereas the LH-response to GnRH is greater than the FSH-response during the reproductive period, in prepubertal children and in menopausal women the FSH response is more pronounced than the LH response, again correlating to the basal levels[38].

3.4.1. GnRH in the Diagnosis of Gonadal Disorders

For the differential diagnosis between primary and secondary hypogonadism the GnRH test is not necessary since the measurement of basal FSH-levels is sufficient for this purpose[186].

Furthermore, the evaluation of the gonadotropin response in patients with radiologically documented pituitary tumours and in those with bonafide hypothalamic lesions including craniopharyngiomas and Kallmann's syndrome has demonstrated that this differential diagnosis cannot be made with the GnRH test. Thus, patients with Kallmann's syndrome may not respond to the first or second test though additional GnRH pulses will finally "wake up" the gonadotrophs and lead to a normal response[190]. In contrast, patients with pituitary tumours and clinical and laboratory evidence of hypogonadism may readily respond to GnRH[59, 186].

In women with severe functional hypothalamic amenorrhea without bleeding after progestagen administration the response to the GnRH test allows one to differentiate three different groups: 1) Normal adult response; 2) FSH response more pronounced than the LH response (prepubertal reaction), 3) Total unresponsiveness[78]. All variants of gonadotropin responses to GnRH can occur in anorexia nervosa with an inverse correlation between the magnitude of the LH increase and the amount of weight loss. However, since in all three situations repetitive administration of GnRH can lead to a normal LH response and chronic, pulsatile GnRH administration to normal ovarian function, the diagnostic value of this test is questionable.

3.4.2. GnRH Test in Hyperprolactinemic Disorders

Though hyperprolactinemia leads to tertiary hypogonadism, the gonadotropin responses to GnRH are variable. Thus, hyper-, normal and blunted gonadotropin responses can be observed[36, 133]. In general, an inverse correlation exists between the duration of amenorrhea and hyperprolactinemia and the height of the LH and FSH rises after GnRH administration[36]. This suggests that long-term suppression of endogenous GnRH pulsatility leads to unresponsiveness of the gonadotrophs similar to the situation in patients with Kallmann's syndrome who have never experienced pulsatility of endogenous GnRH[109]. This is in agreement with the observation that long-term normalization of prolactin levels in these patients restores endogenous GnRH pulsatility and normalizes the GnRH test[133]. Furthermore, patients with long-term hyperprolactinemia who do not respond initially to several pulses of GnRH may still later respond to pulsatile GnRH administration[187], which eventually leads, after two or three days, to normal gonadotropin rises (Fig. 5).

3.4.3. GnRH in Acromegaly

GnRH injection into patients with active acromegaly can lead to remarkable rises of GH, though this is observed less frequently than the inappropriate GH rise after TRH. Thus, only 15% of the patients with active acromegaly will have a GH rise after 100 µg GnRH i.v.[188]. Furthermore, the GH response to GnRH, in contrast to the one seen after TRH, is not generally associated with the paradoxical decrease of GH levels after administration of dopamine agonists[84]. It seems that the majority of acromegalic patients who respond to TRH show no response to GnRH and vice versa[188]. Since this is an infrequent observation with no therapeutic consequence, the GnRH test with subsequent GH evaluation has no place in the routine work-up of acromegalic patients.

3.4.4. Treatment with GnRH

Normal gonadal function in men and women depends on the pulsatile secretion of LH which reflects the pulsatile stimulation of the gonadotroph by hypothalamic GnRH. Leyendecker and coworkers[79] were the first to

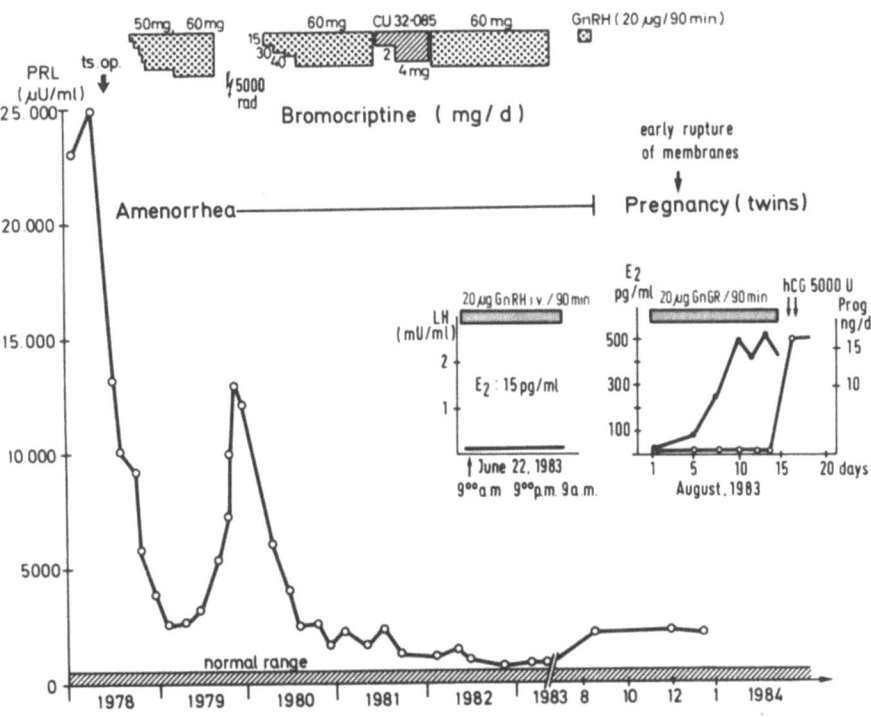

Fig. 5. Prolactin levels and ovarian function in a 31-year-old patient with macroprolactinoma after transsphenoidal surgery, radiotherapy, and long-term dopamine-agonist treatment. Since prolactin levels could not be normalized with various forms of treatment directed against the prolactinoma, pulsatile administration of GnRH was performed to restore gonadal function. Though GnRH treatment in a dosage of 20 µg iv every 90 minutes did not lead to pulsatile LH secretion in the presence of low estrogens (June 22, 1983), long-term pulsatile treatment (August, 1983) led to an increase of estradiol and normal ovulation documented by subsequent twin pregnancy. (With permission from von Werder, 1985)

show that female patients with hypothalamic amenorrhea can be effectively treated with GnRH pulses given *i.v.* at 90 minute intervals. This treatment leads to normal follicular maturation and a high pregnancy rate [78, 79, 100], and is superior to gonadotropin treatment, since severe complications due to overstimulation do not occur. Usually, GnRH is given in a dosage of 5 to 20 µg per pulse every 90 minutes *i.v.* through a tubing system coming from the reservoir of a pump which is programmed accordingly. When ovulation has occurred, the corpus luteum can be maintained either with GnRH or hCG [8]. This treatment has been shown to be effective also in patients with hyperprolactinemia who did not tolerate dopamine agonists[8] or could not achieve normal PRL levels under treatment[187] (Fig. 5).

In young boys GnRH has been used for the treatment of cryptorchidism instead of human chorionic gonadotropin (hCG), though there is no evidence that GnRH is superior to hCG.

Pulsatile GnRH treatment is also effective in males with hypothalamic hypogonadism (Fig. 6). However, GnRH-therapy in the male is less clearly established for the following reasons:

1. To achieve mature spermatozoa in previously untreated patients with hypogonadotropic hypogonadism, three months of treatment are necessary as compared to the two weeks which are needed for follicular maturation in the female[78].

2. For this reason GnRH must be given *s.c.* [54, 190]. In addition, it is more difficult to find the correct GnRH dosage which maintains LH pulsatility over a long period of time[54].

3. A high percentage of males with hypogonadotropic hypogonadism have additional testicular damage, particularly if they have had an operation for undescended testes. Therefore, documentation of normal testicular function by measuring testosterone secretion after hCG is mandatory before treating males with pulsatile GnRH application.

4. Up to now there is no clearcut evidence that treatment with GnRH is superior to treatment with gonadotropins in the male, since problems of overstimulation which occur with gonadotropins in the female are not encountered in the male. Thus, if there is a place for pulsatile GnRH treatment in male hypothalamic hypogonadism or in cases with "slow pulsing" oligospermia, this remains to be established. Although the development of antibodies to GnRH has been reported during chronic treatment with GnRH[178] this adverse effect has not been observed with pulsatile GnRH treatment.

3.4.5. Superactive GnRH Agonists

Administration of superactive GnRH analogues, decapeptides with amino acid substitutions (D-Trp6/GnRH) or nonapeptide-ethyl-amide agonists (Buserelin, Hoechst AG), has a dual effect on gonadotrophs[139]. Initially, they lead to dose-dependent LH and FSH responses which become blunted after repetitive stimulation[139]. Therefore, the indication for the superactive GnRH analogues are situations in which gonadal activity needs to be suppressed rather than stimulated. Chronic treatment with compounds like Buserelin which can be given intranasally, leads to complete inactivation of the pituitary gonadal system. Thus, these agonists have been used successfully as female and male contraceptives[108], for the therapy of endometriosis[180], and for the treatment of true precocious puberty[90].

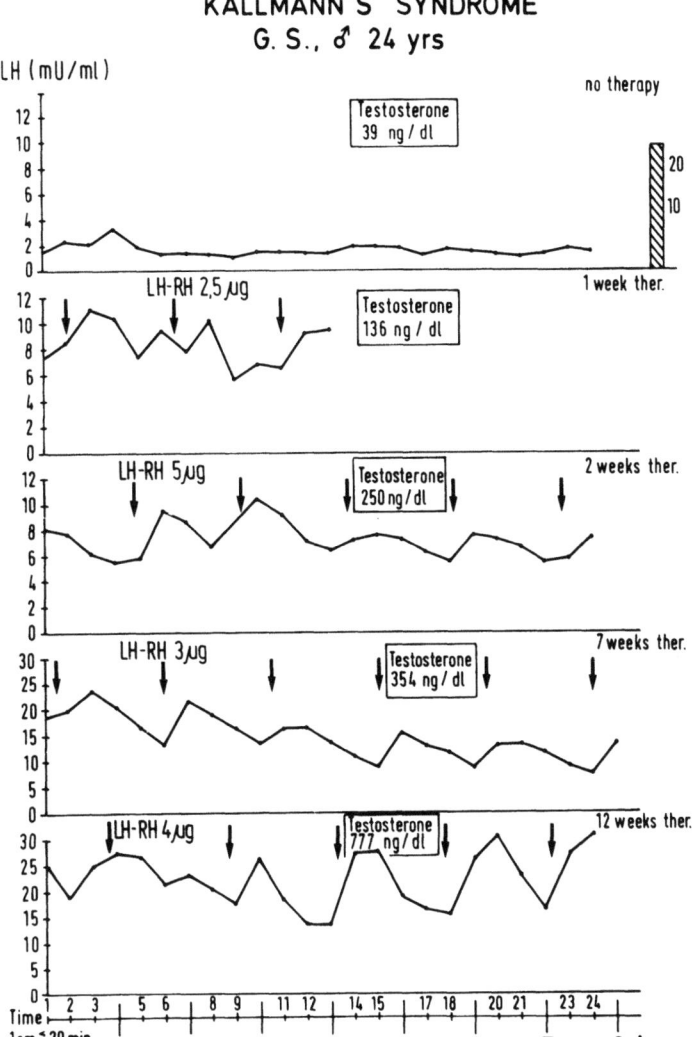

Fig. 6. LH and testosterone levels in a male patient with Kallmann's syndrome (24 years) before and during long-term treatment with sc pulsatile GnRH administration. The rise of testosterone within the normal range was accompanied by testicular growth up to 12 ml. (With permission from von Werder and Müller, 1985)

Furthermore, GnRH analogues have been used for treatment of hormone dependent cancer. In patients with metastatic prostatic cancer, GnRH analogues are as effective as orchiectomy and have become an alternative to surgical castration[71]. Since these peptides initially lead to stimulation of gonadotropin secretion, (see above) an early "flare up" of the metastatic prostatic cancer has been observed. For this reason, the combination of GnRH superactive agonist with antiandrogens has been recommended[17], particularly since they also inhibit the effect of adrenal androgens, which are not influenced by the superactive agonists.

Also GnRH analogues have been used in premenopausal breast cancer patients with beneficial results, though this treatment ist still under clinical investigation[180].

4. Corticotropin Releasing Hormone (CRH)

Though it was quite evident, that the posterior lobe peptide vasopressin, which also stimulates ACTH-secretion[123], could not be the main factor stimulating ACTH-release, the specific corticotropin releasing factor remained elusive for a long time. It was therefore a great breakthrough when Vale and his coworkers reported the isolation, sequence, synthesis, and biological activity of a synthetic replicate of a 41 amino acid ovine hypothalamic corticotropin releasing factor in 1981[174].

That it took so long to elucidate the structure of CRH is due to several factors. The ACTH-assay is difficult and therefore the bioassay of hypothalamic extracts for CRF-activity. Furthermore, most hypothalamic extracts contain already ACTH, which interfers with the bioassay. One of the main obstacles was that even after purification, due to similar molecular size (ACTH with 39 amino acids versus CRH with 41 amino acids, Table 1) the CRH bioactivity after gel chromatography was overlooked because the ACTH generated was blamed on the ACTH contamination of the purified extract[173]. Only picomol levels of CRH are present in each hypothalamic fragment necessitating a great number of hypothalamic fragments in addition to a good yield during purification[173, 174].

4.1. Structure of CRH

The ovine CRH molecule is closely homologous to two other naturally occurring peptides, sauvagine, isolated from frog skin, and urotensin-I, isolated from the urophysis of certain fishes[173]. In contrast to the shorter hypophysiotropic peptides TRH, GnRH, and SRIF, CRH like GRH is structurally species specific, though there seems to be no species specificity in respect to the biological activity[173]. In addition to ovine CRH the structure of porcine[173] and rat CRH has been elucidated[173]. Furutani et al.[39] and Shibahara and coworkers[151] elucidated the nucleotide sequence of the genes for both ovine and human prepro-CRH and deduced the amino sequence of the corresponding CRH-peptides, which are amidated at the C-terminus. The human CRH differs from the ovine molecule in seven amino acids at positions 2, 22, 23, 25, 38, 39, and 41[173]. Human and rat CRH are chemically identical[151]. Most of the amino acid changes are localized in the C-terminal part of the molecule which seems to be crucial for the biological activity. Thus, deletion of the C-terminal amino acids completely abolishes the bioactivity of CRH. Already deamidation of the C-terminus leads to reduction of the biological activity to 0.1%[131].

Deletion of N-terminal amino acids leads to peptides which act as competitive inhibitors of CRH *in vitro*. Substitution of amino acids leading to an optimal alpha-helical structure of the CRH molecule results in peptides which are twice as potent as the parent CRH peptide in an *in vitro* model[131].

4.2. Distribution of CRH

CRH-like immunoreactivity has been detected by radioimmunoassay in extracts of the stalk median eminence region of several species. Immunostaining allowed to identify CRH-immunoreactive cell bodies in the paraventricular nucleus. Furthermore, CRH has been detected in other hypothalamic regions including the periventricular area and the supraoptic nuclei. They seem to project into the external zone of the median eminence, where CRH is released into the capillaries of the portal system.

Most of the CRH stained cells are only detectable if intact animals are pretreated with colchicine[173]. Co-localization of CRH with other neuropeptides has been observed. In the paraventricular nucleus, neurotensin and oxytocin were found together with CRH in the same neurons. The OT/CRH neurons project into the posterior pituitary whereas the neurotensin/CRH neurons project into the median eminence[173].

CRH-like immunoreactive material has been demonstrated in various extrahypothalamic sites within the CNS as well as in the gastrointestinal tract, the adrenal medulla and the pancreas[173]. Thus, the extrapituitary actions of corticotropin releasing hormone include a CNS effect with stimulation of general locomotor activity, a cardiovascular effect, particularly hypotension, and an inhibitory effect on insulin release from the perfused rat pancreas[173].

4.3. Measurement of CRH

In agreement with the hypophysiotropic role of CRH, CRH immunoreactivity could be detected in plasma extracts from portal blood in rats whereas it was undetectable in the peripheral rat blood, extracted and assayed in the same way[173]. Using a human homologous CRH radioimmunoassay and an immunoaffinity procedure Suda and coworkers were able to measure CRH concentrations in normal subjects[165]. The plasma immunoreactive CRH levels were $6 \pm 0.5 \, pg/ml \pm SD$ and showed a diurnal variation in the same fashion as ACTH and cortisol. CRH levels increased during stress situations and could be suppressed by administration of glucocorticoids, again in parallel fashion to circulating ACTH and cortisol levels, suggesting that the immunoreactivity measured in the periphery reflects changes of CRH concentration in portal blood[165]. The same investigators also demonstrated high CRH levels in patients with Addison's disease, Nelson's syndrome and, interestingly, in patients with hypopituitarism due to compression of the pituitary by an intrasellar tumour[165]. CRH immunoreactivity was very low in patients with Cushing's syndrome and also in corticosteroid treated patients. Similar findings were made by Stalla and coworkers, suggesting that in the future measurement of CRH immunoreactivity may even be helpful for making the differential diagnosis between hypothalamic and pituitary pathology[163, 164]. However, Charlton et al. measuring CRH in plasma of normal subjects and depressed patients with a sensitive immunoradiometric assay, found no significant difference between the CRH levels of normals and depressed patients and no correlation between CRH- and cortisol-levels, which tended to be higher in the depressed patients[21]. Since CRH is also produced in the gastrointestinal tract, an extrahypothalamic source for CRH in peripheral blood therefore cannot be ruled out[21]. The latter is definitely the case in pregnant females, in whom elevated CRH-levels are measured particularly in the 3. trimester[164]. This immunoreactive CRH, for which regular bioactivity has been demonstrated, is of placental origin[163]. It is not clear why pregnant females do not have higher ACTH levels due to the elevated biologically active CRH-levels. One explanation would be that the minor elevation of free cortisol occurring in pregnant females inhibits a further rise of ACTH, due to feedback inhibition at the pituitary level[42, 123].

4.4. CRH as a Hypophysiotropic Hormone

CRH plays an important role in the stress response. The latter was termed by Selye as a "general adaptation syndrome", which is found throughout the animal kingdom[147]. The response occurs if a stressor threatens metabolic homeostasis. The reaction to this noxious stimulus is mediated by the brain and manifested as behavioral as well as cardiovascular and metabolic changes. Some of these effects are reproducible by intraventricular administration of CRH, demonstrating the neurotransmitter role of the hyophysiotropic hormone[173]. However, the most important effect in this adaptative response is probably the activation of the pituitary adrenal axis.

CRH stimulates the release of all proopiomelanocortin (POMC) derived peptides, ACTH, β-endorphin, and β-lipotropin[174]. This effect is mediated by specific CRH receptors at the corticotrophs which are like the GRH receptors, coupled to the adenyl cyclase system[42]. Thus, CRH leads to a dose dependent accumulation of cAMP and addition of theophylline, an inhibitor of cyclic nucleotide phosphodiesterase stimulates ACTH-release in pituitary cell cultures[72]. The CRH-induced ACTH secretion is also calcium dependent since low concentrations of calcium in the medium attenuate the ACTH response to CRH[123].

Acutely, CRH leads to release of stored POMC-peptides with a decrease of intracellular ACTH[173]. However, CRH also stimulates ACTH-synthesis demonstrated by an increase of the POMC messenger RNA[149, 173].

However, there are several other factors, which lead to stimulation of ACTH secretion. Thus, vasopressin is a well known factor, which has been used clinically for stimulation of ACTH secretion in humans[91]. It has been shown that vasopressin acts synergistically to CRH, potentiating the ACTH-response to CRH[43, 74, 132]. Activation of alpha-receptors also seems to have a stimulatory effect on ACTH secretion[123]. This observation explains the finding, that even the highest dosages of CRH never lead to such ACTH increases as they are observed during insulin hypoglycemia in humans[102, 103]. In this situation the stress response is obviously not only mediated by CRH but by additional factors like vasopressin for example. In rat and in human fetus, POMC derived peptides are also secreted from the intermediate lobe which is under different hypothalamic control and shows different processing of the POMC precursor peptide[173]. Again, CRH stimulates the secretion of POMC-peptides which is inhibited by dopamine in a non competitive fashion[173]. In adult humans CRH-induced ACTH-secretion is effectively blocked by corticosteroids, demonstrating that in addition to the feedback inhibition at the hypothalamic level, corticosteroids exert an acute inhibitory effect at the pituitary level[69].

4.5. Pathophysiology of CRH

Secondary adrenal insufficiency can be due to a hypothalamic CRH-deficiency. The latter can be idiopathic without any demonstrable lesion in the pituitary stalk or the suprasellar area or due to inflammatory or expanding lesions in this particular region[102, 191]. An isolated CRH-deficiency occurs rarely, whereas hypothalamic adrenal insufficiency together with other anterior pituitary deficiencies due to a suprasellar tumor may be observed more frequently[191]. That ACTH hypersecretion leading to Cushing's disease is due to excessive CRH secretion can rarely be proven[69, 111]. Only in the rare ectopic CRH syndrome, CRH-dependent Cushing's disease is proven. However, only very few cases with this pathophysiology have been reported[20]. The latter is documented only when CRH can be measured in tumour extract and in the peripheral circulation and there is eutopic ACTH hypersecretion from the anterior pituitary[20, 24]. In the majority of patients with Cushing's disease the ACTH excess seems to stem from a pituitary microcorticotrophinoma, independent of hypothalamic CRH. This is suggested by the high efficacy of microsurgical extirpation of the microadenomas leading to clinical and biochemical remission of

Cushing's disease[69]. However, some patients may have rapid recurrence after presumed selective adenoma removal and in other patients no microadenoma may be detected[69, 104, 111]. If ACTH hypersecretion in these patients is due to endogenous CRH hypersecretion is not proven. Recently, it has been suggested, that hyperpulsatility of cortisol in these patients may reflect the pulsatile activity of CRH neurons, which cannot be detected in patients with pituitary corticotrophinomas or adrenal cortisol producing tumors[176].

4.6. Clinical Utilization of CRH

Four years ago, ovine and, more recently, human CRH have become available for investigating pituitary function in normal subjects and in those with hypothalamic pituitary diseases[102]. In normal subjects i.v. administration of ovine and human CRH leads to prompt activation of the pituitary adrenal axis with ACTH peak levels occurring within 30 minutes and peak cortisol levels between 45 and 60 minutes after CRH administration[102]. Since the half life of ovine CRH is longer than that of hCRH, the ACTH response after oCRH is more prolonged than after hCRH[164].

However, for theoretical reasons, the homologous peptide should be used for diagnostic purposes in humans[164]. There is a dose response relationship between CRH and the ACTH response, though there is a wide variability of ACTH and cortisol responses to the same dosages in different subjects[112]. In contrast to GRH, CRH infusion leads to sustained elevation of ACTH secretion[164]. However, this observation is not important for diagnostic purposes. Though serious side effects occur rarely with dosages below 100 µg, some subjects complain of flushing, headaches and neck oppression[102, 103]. High dosages can cause cardiovascular side effects, particularly low blood pressure and in rare cases shock symptoms[144].

4.6.1. Use of CRH in the Differential Diagnosis of Adrenal Failure

Evaluation of CRH-induced ACTH secretion is not necessary for making the differential diagnosis between primary and secondary adrenal failure. However, this test may be useful in differentiating between hypothalamic- and pituitary-related adrenal insufficiency[102, 191]. Whereas a prompt rise of ACTH from low basal levels is usually observed after CRH administration in patients with hypothalamic failure, ACTH levels remain undetectable in patients with pituitary damage (Fig. 7). Cortisol levels remain unchanged due to adrenal unresponsiveness to endogenous as well as exogenous ACTH because of structural atrophy of the adrenal cortex.

Furthermore, the CRH test complements the GRH test for localizing a hypothalamic or pituitary lesion, when performed in combination with the

The Biological Role

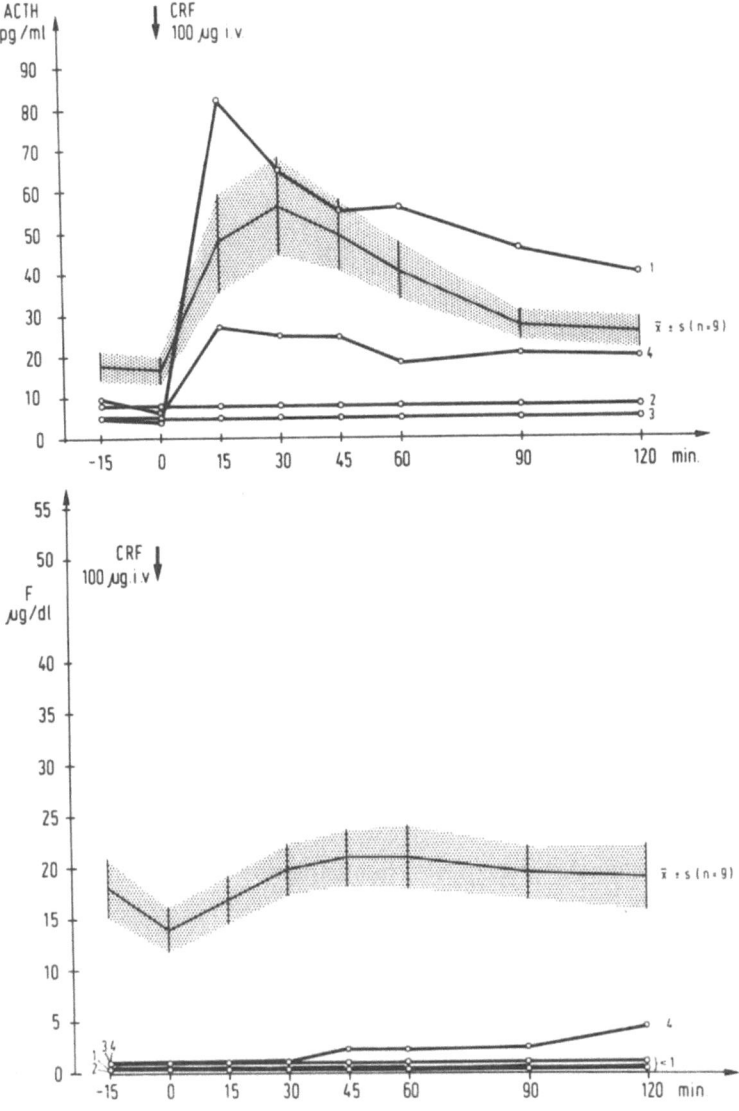

Fig. 7. Patterns of ACTH and cortisol secretion after CRH administration in 4 patients with secondary adrenal failure without radiological findings suggesting intra- or suprasellar lesions. Cortisol levels did not increase in any of the patients, but various ACTH responses were observed. In two patients with isolated ACTH deficiency, no change of the unmeasurable ACTH levels were seen (2, 3) suggesting a pituitary defect. In the other two patients ACTH levels increased from low baseline to normal levels, reaching a maximum 15 minutes after CRH (1, 4) indicating a hypothalamic CRH deficiency. Normal ACTH and cortisol responses are shown by the shaded area. (With permission from Müller et al., 1982)

insulin-induced hypoglycemia test, which leads to activation of GH and ACTH secretion in the presence of a functioning hypothalamus (Fig. 12).

CRH stimulates only the secretion of ACTH or POMC derived peptides. This is also the case in active acromegaly where no inappropriate GH secretion has been observed after CRH[191].

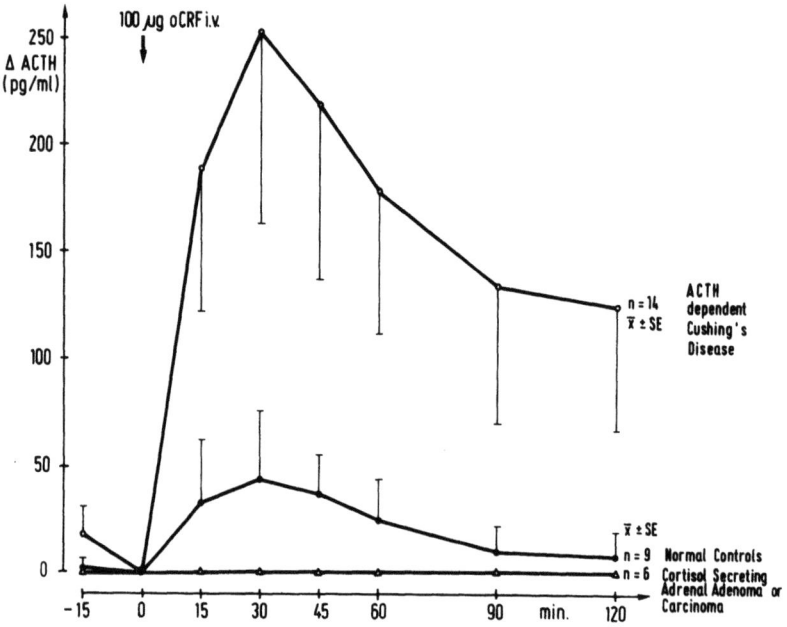

Fig. 8. ACTH response after CRH stimulation in 14 patients with ACTH-dependent Cushing's disease and in 6 patients with autonomous cortisol secretion due to an adrenal adenoma or carcinoma. The ACTH response in normal controls is also shown. For better comparison the absolute ACTH increases in pg/ml are shown. (With permission from Müller et al., 1985)

4.6.2. Use of CRH as a Diagnostic Aid in Cushing's Syndrome

Due to the wide variation of ACTH rises after CRH, evaluation of CRH-induced ACTH and cortisol secretion is not suitable for making the diagnosis of Cushing's syndrome[24, 104]. However, this test has been shown to be useful for the differential diagnosis of Cushing's syndrome. Thus, patients with pituitary ACTH hypersecretion and bilateral adrenal hyperplasia usually show dramatic responses of ACTH and cortisol levels after CRH, whereas patients with unilateral adrenal cortisol-secreting adenomas show no such rise and their ACTH levels remain suppressed (Fig. 8,[104, 120]). Furthermore, patients with ectopic ACTH secretion show in general no significant increase after CRH of the generally very elevated ACTH levels[104], though rare exceptions have been reported.

After successful microadenomectomy in Cushing's disease, basal levels of ACTH, which have decreased to below normal or within the normal range, cannot be stimulated by a single bolus of CRH[191]. Thus, postoperative corticotroph insufficiency apparently indicates a successful microadenomectomy.

Very rarely pituitary hypersecretion of ACTH causing Cushing's syndrome is due to ectopic CRH production. Recently, such a case has been reported in whom the diagnosis has been made by measuring CRH immunoreactivity[20]. Although in this patient the diagnosis had been made postmortem, CRH measurements should exclude this rare condition in patients with Cushing's syndrome in the future.

In patients in whom an adrenal cortisol-secreting tumour has been removed, contralateral secondary adrenal insufficiency may persist for a long period of time. However, repetitive administration of CRH can "wake up" inactive corticotrophs, thus demonstrating that, in addition to the negative feedback of corticosteroids at the pituitary level, the main reason for persisting adrenal insufficiency is the failure of the CRH neurons to resume activity[191].

Up to now there is no evidence that CRH may also have a therapeutic value in addition to its diagnostic usefulness.

5. Growth Hormone Releasing Hormone (GRH)

Reichlin postulated the existence of a growth hormone releasing factor already in 1960[127]. Four years later Deuben and Meites demonstrated GRF-activity in extracts of rat hypothalami[26]. However, purification, isolation and elucidation of the structure of this hypothalamic releasing hormone was achieved only in 1982 when Guillemin and coworkers[49] isolated a peptide with GRF-activity from an islet cell carcinoma of an acromegalic patient (human pancreatic growth hormone releasing factor = hpGRF). The synthetic copy of this peptide with 44 amino acids stimulated GH-secretion *in vivo* and *in vitro*. At the same time Thorner and coworkers extracted a GRF-molecule with 40 amino acids from a benign islet cell adenoma which had also led to acromegaly[169]. The 40 N-terminal amino acids of this GRF-peptide were identical with the hpGRF 1–44[50].

5.1. Structure and Biological Activity of GRH

Guillemin found three GRF-peptides in his endocrine active pancreatic carcinoma, hpGRF 1–44, hpGRF 1–40, and hpGRF 1–37, which differed only in the C-terminal region by deletion of 4 and 7 amino acids respectively[49] (Table 1). hpGRF 1–40 was the only GRF peptide extracted from the tumour of the acromegalic patient by Thorner[169]. The product of the GRH-gene expression is GRH 1–44 with the amidated carboxyl terminus[47]. This peptide was found in human hypothalamus as well as GRH 1–40 with a free carboxyl terminal[82, 83]. It is therefore justified to skip the prefix human pancreatic (hp) and use the term growth hormone releasing hormone GRH, though Guillemin and corworker have suggested the name somatocrinin for GRH in analogy to somatostatin[50].

GRH peptides from other species (porcine, ovine, caprine) show significant homology to human GRH. The only changes occur in the C-terminal amino acid sequence which is not necessary for the biological activity of the molecule. An exception is the rat which has GRH-molecule with only 43 amino acids[83].

The structure of the gene for human prepro-GRH is known, it codes a single chain peptide with 108 amino acids with an N-terminal signal peptide of 20 and a pro part of 12 amino acids. GRH 1–44 is positioned between the 32. and 76. amino acid followed by C-terminal part of 23 amino acids. During intragranular processing GRH 1–44 is cleaved at the position of the basic amino acids and the C-terminus is amidated[47].

In contrast to CRH, deletion of the C-terminal fragments does not lead to loss of biological activity, Thus GRH 1–29 has the same biological activity as GRH 1–40 and GRH 1–44 *in vivo* (Fig. 9). Only when further amino acids at the C-terminus are deleted, does the neurohormone lose its biological activity[50, 88].

Deletion of N-terminal amino acids leads immediately to loss of biological activity. Furthermore, GRH 4–44 has a GRH antagonistic activity[50]. In contrast, substitution of aminoacids of the 29 residue peptide may lead to superactive GRH-agonists[75].

In contrast to the *in vivo* situation, *in vitro* investigations on monolayer rat pituitary cell cultures showed that the C-terminal amidated form of GRH 1–44 has a higher biological activity, compared to the C-terminal shortened GRH-peptides like GRH 1–40 and GRH 1–37[50]. The biological efficacy, *i.e.* secretion of immunoreactive rat GH can be shown within 30 seconds and runs parallel to the increase of cyclic AMP in the medium. GRH stimulates the exocytosis of GH containing granules and the activation of the cyclic AMP dependent protein kinases via specific receptors localized in somatotroph granules in a dose-dependent fashion. Thus, GRH could be identified in secretory GH-granules of human somatotroph cells[50]. Increasing dosages of cyclic AMP lead to the same dose response curves as obtained with increasing dosages of GRH. In contrast, other second messengers like prostaglandin E2, which also stimulates GH-secretion lead to a different dose response relationship. Accordingly, prostaglandins do not stimulate GH-secretion via the GRH receptor[50].

Somatostatin does not inhibit GRH stimulated GH-secretion in a competitive fashion but via its own specific somatostatin receptor[77]. Furthermore, somatomedin C or insulin like growth factor (IGF I) inhibits

GRH-stimulated GH-secretion *in vitro* whereas IGF II shows negligible inhibitory effects[2, 50]. In agreement with the rapid effect of GRH on GH-secretion, *i.e.* stimulation of exocytosis of already synthesized ready available GH, protein synthesis blockers like cyclohexemide have no effect on GRH-stimulated GH-secretion[50]. However, GRH also stimulates GH synthesis, which can be demonstrated by the rapid increase of GH messenger RNA in pituitary cells after addition of GRH. This effect is highly specific, since the messenger RNA for other anterior pituitary hormones is not influenced[50].

There is little evidence that GRH has other hypophysiotropic effects but stimulating GH-secretion from somatotroph cells. In high concentrations, particularly in patients with acromegaly GRH can stimulate PRL secretion[186]. The only extrapituitary biological activity demonstrated sofar is that GRH can stimulate exocrine pancreatic function[114].

5.2. Distribution of GRH

The highest concentrations of GRH are found in the hypothalamus particularly in the mediobasal part and in the nucleus arcuatus, near the infundibulum[12]. Furthermore, GRH positive nerve fibers are found in the median eminence where the capillaries of the portal system are reached. GRH-positive neurons project into the anterior hypothalamus and into the nucleus paraventricularis. Since GRH-positive neurons end at nerve cell bodies it is possible that GRH, in addition to its neurohumoral role, plays also a neurotransmitter role[50]. This is particularly probable since GRH has been demonstrated outside the hypothalamus in other parts of the CNS[12].

GRH is found outside the CNS in the intestinal tract, particularly in the duodenum and jejunum[153]. Chromatography of gut extracts demonstrates in addition to monomeric GRH a higher molecular GRH peptide[153]. Like all other releasing hormones[157], it is also found in the placenta[3].

5.3. Measurement of GRH

The measurement of GRH-activity was initially performed using pituitary cell cultures *in vitro*, which allowed measurement of GRH in picomol concentrations[50]. After elucidation of its structure, radioimmunoassays have been developed with a lower limit of detection down to 10 pg GRH/ml. So far, divergent results have been reported concerning GRH levels in normal subjects in the peripheral circulation. Saito *et al.* found a mean immunoreactive GRH plasma concentration of 10.3 ± 4.1 pg/ml in normal subjects[137]. Besser and coworkers in London reported a GRH radioimmunoassay with a lower limit of detection of 10 pg/ml[117]. With this radioimmunoassay they found measurable GRH levels in 75% of 37 healthy normal subjects in the range between 10 and

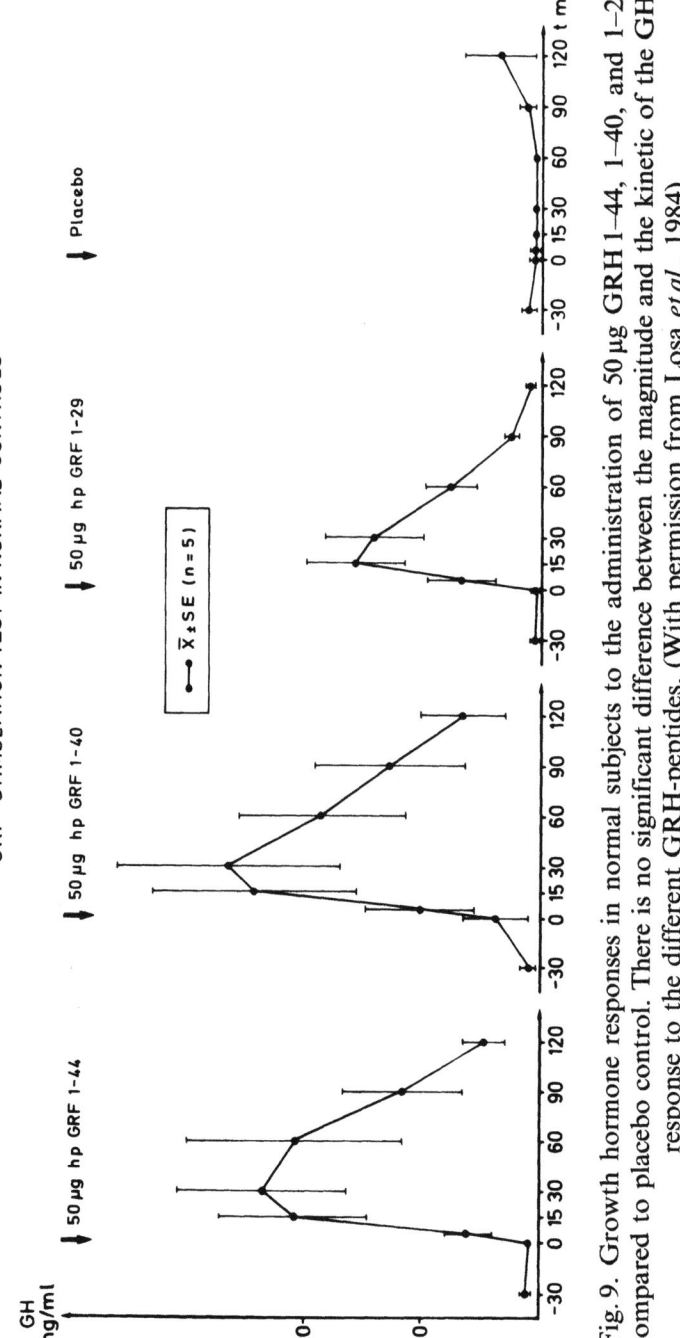

Fig. 9. Growth hormone responses in normal subjects to the administration of 50 μg GRH 1-44, 1-40, and 1-29 compared to placebo control. There is no significant difference between the magnitude and the kinetic of the GH-response to the different GRH-peptides. (With permission from Losa et al., 1984)

50 pg/ml. Other authors have failed to demonstrate measurable GRH-levels in normal subjects[89].

In respect to the low concentrations of GRH in the human hypothalamus it is doubtful that hypothalamic GRH can be measured in peripheral circulation. Most likely, the latter stems from the intestinal tract which is further confirmed by the fact that GRH immunoreactivity in the peripheral circulation seems to increase with food intake and does not correlate to circulating GH-levels[162].

The half time of disappearance of GRH after bolus injection of GRH is 7.5 min (first phase,[89]) and between 40 and 50 minutes for the second phase[181].

5.4. Clinical Utilization of GRH

5.4.1. Biological Activity of GRH in Normal Subjects

The intravenous administration of GRH leads to a prompt rise of GH-levels in normal subjects (Fig. 9). Other hormones are not influenced[171]. There are no side effects except short lasting flush symptoms and minor neck oppression[87, 88, 89, 192]. There is a dose response relationship between injected GRH and the GH response in the low dose range. Thus, 12.5 µg GRH 1–44 leads to a significant increase of the GH-levels and the maximal GH-secretion is reached with 50 µg GRH (Fig. 10). Further increases of the dosage up to 10 µg/kg body weight or 600 µg GRH 1–44 do not lead to a further increase of the GH-levels. The GRH induced GH-responses vary considerably in different individuals, though the maximal GH-level is usually reached 15–30 min after GRH-injection[88, 89]. The individual GH-responses are most likely due to the different endogenous somatostatin levels in the portal blood which are known to have a modulating influence on GRH stimulated GH-secretion. Furthermore, passive immunization with somatostatin antibodies in rats demonstrated that the spontaneous pulsatile GH-secretion is not due to intermittent hypothalamic release of somatostatin but to the pulsatile activity of the GRH neurons. Accordingly, monoclonal GRH antibodies suppressed pulsatile GH-secretion in rats[50].

Repeated bolus injection of GRH at two hour intervals does not lead to an increase of the GH response but to a blunting of GH secretion[85]. Continuous infusion of GRH does not lead to a further increase of the GH-levels but to a decrease, despite continuously elevated GRH blood levels (Fig. 11).

Since similar findings were made *in vitro*, a counterregulatory effect of somatostatin as a cause of this phenomenon is unlikely[85]. It has been demonstrated that these findings are partly due to desensitization and partly due to depletion of the ready releasable GH-pool of the somatotrophic cells[86].

If GRH is given at six hour intervals, regular GH-increases are again observed. When GRH is infused into normal human subjects over 24 hrs in a low dosage an augmented pulsatile GH release is observed compared to a

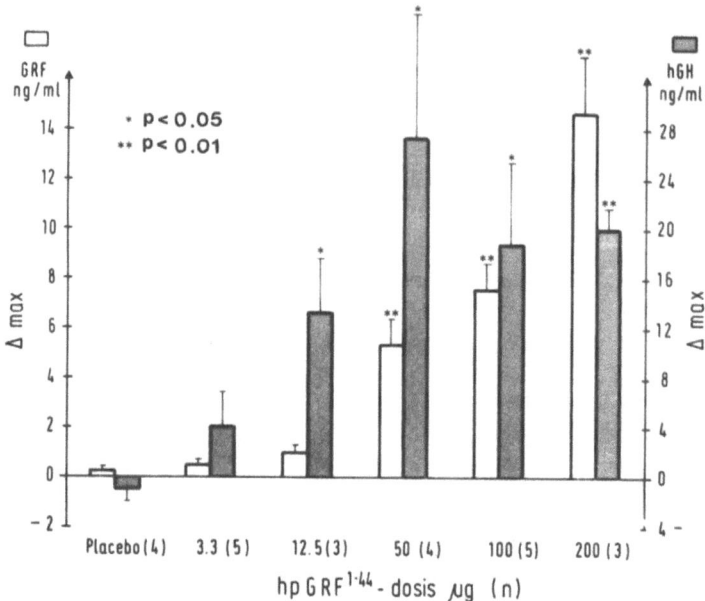

Fig. 10. Maximal increases of GRH 1–44 and GH levels after intravenous administration of synthetic GRH 1–44 to 8 normal volunteers. The maximal increase during the test was calculated for each individual subject. The height of the bars represent the mean of the individual values ($\pm$ SE). There is an obvious dose response relationship between the GRH immunoreactivity measured in plasma and the dosages of GRH administered. However, a dose related GH-response is only observed up to 50 µg GRH 1–44 though the increase after 12.5 µg GRH is already not statistically significant from the responses after the administration of 50, 100, or 200 µg GRH. (With permission from Losa et al., 1983)

saline infusion on a control day[177]. This finding could be due to GRH induced suppression of somatostatin release. When acute stimulation of GH secretion with GRH is followed by an insulin hypoglycemia test (IHT), a regular rise of GH can be observed during the IHT suggesting that the IHT induced GH response is not mediated by GRH[152]. The GRH induced rise of GH in humans is significantly suppressed by corticoids[175], elevated blood sugar levels[148], elevated free fatty acids[57], in obese patients[194], and by certain drugs, for example theophylline [Losa M, et al. (1986) Acta Endocr 112: 473–480].

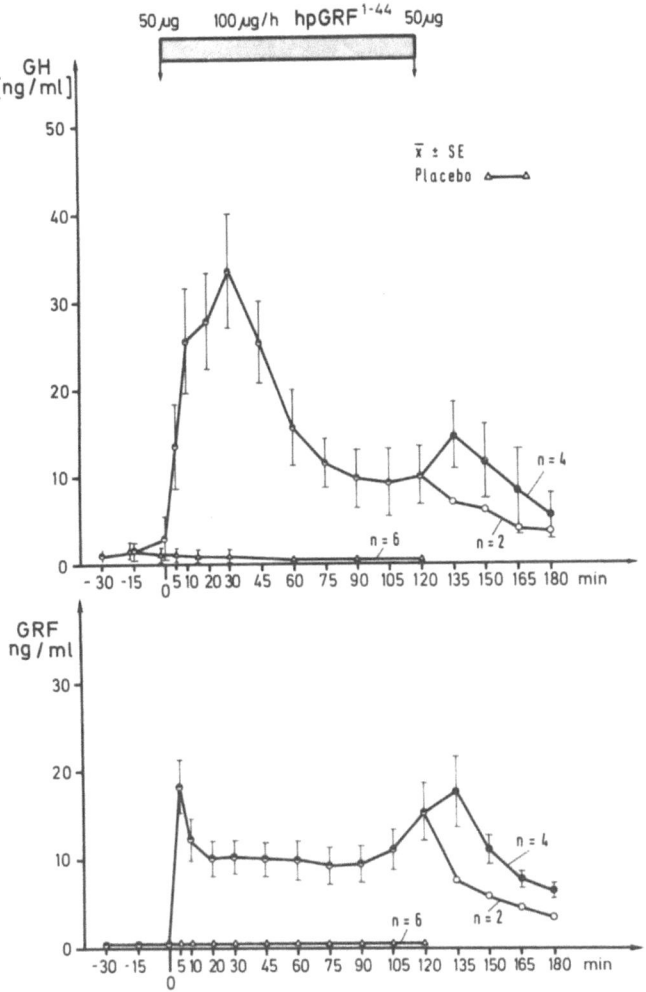

Fig. 11. GH-levels and plasma GRH immunoreactivity in 6 normal subjects after 50 µg GRH 1–44 bolus injection followed by infusion of 100 µg GRH/h. After 2 h of infusion 4 subjects were given another 50 µg GRH-bolus whereas the other 2 received 0.9% saline only. In a control test only saline was given. During this experiment GRH could not be detected at any time. Though continuous GRH-infusion leads to continuously elevated GRH-levels, GH-levels decreased and only a blunted response to the second GRH-bolus is observed. (With permission from Losa et al., 1984)

5.4.2. Evaluation of Anterior Pituitary Function with GRH

Since GH secretion is probably the anterior pituitary function most sensitive to hypothalamic or pituitary damage, evaluation of GH-secretion in patients with suspected hypothalamic and/or pituitary disease is of

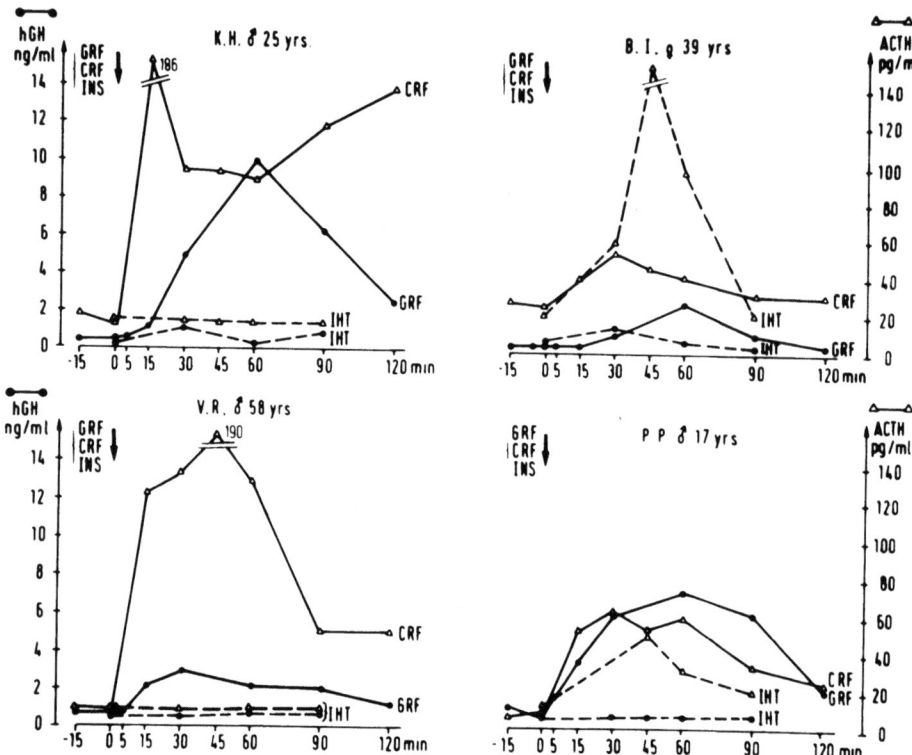

Fig. 12. Stimulation of growth hormone (GH, ●) and ACTH (△) secretion during insulin-induced hypoglycemia test (IHT, -----) and after iv administration of the releasing hormones GRH and CRH (——) in 4 patients with suprasellar tumours and partial or complete anterior pituitary failure. The normal increase of GH and ACTH after administration of the releasing hormones is compatible with a primary hypothalamic lesion and a normally functioning anterior pituitary lobe. Both patients B. I. and patient P. P. (right side) had suprasellar craniopharyngiomas with partially impaired GH secretion (no stimulation during IHT) but normal ACTH secretion during IHT and after CRH. This demonstrates again that the corticotrophic function is more resistant to damage than the somatotrophic one. (With permission from von Werder and Müller, 1985)

diagnostic value. Since GRH stimulates the somatotroph directly, whereas, for example, insulin-induced hypoglycemia needs an intact hypothalamus to induce GH secretion, a combination of both tests may allow one to localize the defect (Fig. 12). A completely normal GH response to GRH in a patient who does not respond to insulin-induced hypoglycemia indicates a suprasellar lesion causing endogenous GRH deficiency (Fig. 12). However, no definite statement about the value of this test can be made yet because experience is still limited[13, 191, 192].

5.4.3. Use of GRH in the Differential Diagnosis and Treatment of Pituitary Dwarfism

The possibility of performing GRH tests in patients with idiopathic GH deficiency has revealed that the majority of these patients show a GH response after one or multiple GRH pulses, thus documenting endogenous GRH deficiency as a frequent cause of pituitary dwarfism[13, 75]. Very often,

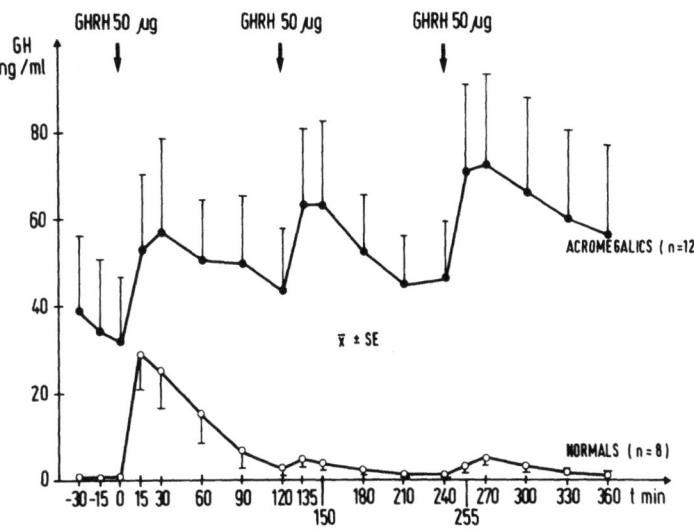

Fig. 13. GH-levels after 3 injections of 50 μg GRH in 2 hours intervals in 12 patients with acromegaly and 8 normal subjects. Whereas in normal subjects the second and the third GRH-bolus leads to a blunted GH-response, clearly distinct and comparable increases of GH after each GRH bolus can be observed in acromegalic patients. (With permission from Losa et al., 1985)

despite modest GH increases, a rise of somatomedin-C levels within 12 hours after GRH is observed, indicating a possible therapeutic role of GRH in these patients[13]. Thus, repetitive *s.c.* administration of 1–3 μg GRH per kg body weight every 3 hours over a period of 6 months has been shown to stimulate growth velocity in a fashion comparable to the one obtained with GH[170].

5.4.4. GRH as a Diagnostic Aid in Acromegaly

The administration of 100 μg GRH to patients with active acromegaly usually leads to a prompt rise of GH, the height of the response correlating often to basal GH levels but not to the activity of the disease[87, 154]. Thus, no or blunted GH responses to GRH have been seen in patients with still active acromegaly, though the majority of these patients had been treated previously[87]. In contrast to normal subjects, repeated injection or infusion of GRH does not lead to decreasing GH-responses of the initially elevated GH-levels, suggesting a larger ready releasable GH pool in these patients (Fig. 13).

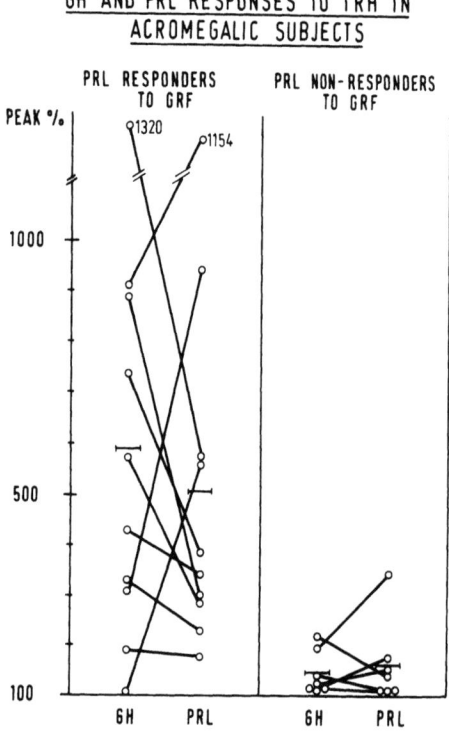

Fig. 14. Maximal prolactin and GH peaks to 200 ug TRH in 17 acromegalics. The patients are divided on the basis of a prolactin response to GRH(F) in PRL responders (n = 10, left panel) and PRL nonresponders (n = 7, right panel). Both GH and PRL-responses to TRH are significantly higher in the PRL-responders than in the PRL non-responders. The horizontal lines represent the mean GH- and PRL-response to TRH for each group. (With permission from Losa et al., 1985)

There is no correlation between the GRH-induced GH rise and the result of other stimulatory or suppressive tests like TRH, GnRH, insulin-induced hypoglycemia, and oral glucose tolerance test[87]. Furthermore, 50% of acromegalics show a rise of PRL after GRH, which is not observed in normal subjects. It is of interest that patients who show a PRL response to GRH also show a GH and PRL response to TRH (Fig. 14). Thus, the regulation of prolactin secretion in these patients is similar to the one of GH and the regulation of GH is similar to the regulation of normal PRL secretion. In agreement with this pattern, these patients also show a paradoxical suppression of GH levels after the administration of dopamine agonists[84]. It is reasonable to speculate that in these instances PRL and growth hormone stem from the same somatomammotrophic cell.

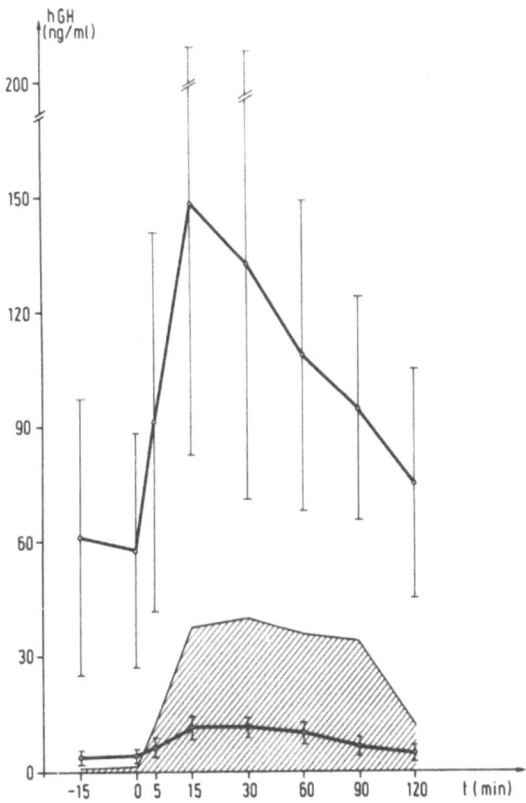

Fig. 15. GRH test with 100 μg GRH iv in 8 acromegalic patients before and after transsphenoidal surgery which led to clinical cure of active acromegaly. After surgery, normalization or a significant fall of the basal GH level and of the GH response to GRH is seen. The shaded area represents the normal range of the GH rise observed in 12 normal controls. (With permission from Losa et al., 1985)

After surgery when basal GH levels have been normalized, the response of GH to GRH is usually normal or subnormal (Fig. 15). Therefore, the result of the GRH test does not provide any relevant pre- or postoperative diagnostic information[87].

Less than 1% of the acromegalic patients harbor ectopic GRH producing tumours[9, 141], which are mostly located in the pancreatic islets, in the lung or in the gut (Fig. 16). Since GRH radioimmunoassays are available, GRH can be measured easily in patients with elevated GRH levels originating from an ectopic source and leading to GH hypersecretion[169, 189]. Therefore a single GRH measurement has been recommended in every acromegalic to exclude the rare syndrome of ectopic GRH hypersecretion[168]. Even rarer than ectopic GRH production is ectopic GH-production, which can be proven by demonstration of GH-messenger RNA in the tumour[97].

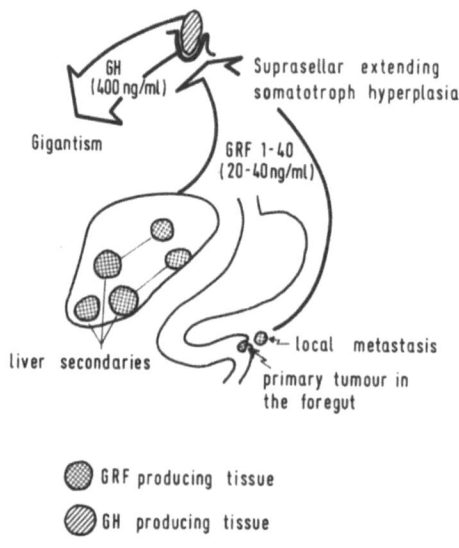

Fig. 16. GRH-producing tumour with a local lymph node metastasis and liver metastases. The high GRH-level in the peripheral circulation has led to a suprasellar extending somatotroph hyperplasia and to gigantism in the 15-year-old patient (von Werder et al., 1985)

6. Somatostatin (SRIF)

While looking for growth hormone releasing activity in Guillemin's laboratory at the Salk Institute, Brazeau and coworkers[15] were able to isolate and sequence a cyclic peptide which was able to inhibit the release of growth hormone secretion. This 14 amino acid peptide (Table 1) was called somatotropin-release inhibitory factor (= SRIF) or somatostatin. It was actually a rediscovery because Krulich et al.[70] had described already in 1968 a GH-release inhibitory substance, which they had extracted from rat hypothalamic fragments. However, biochemical and clinical investigations concerning the physiology and pathophysiology of somatostatin started only after 1973 leading to a rapid expansion up to our present knowledge. Several reviews have appeared recently in which the accumulating literature about this peptide has been digested[124, 126].

It has become clear that somatostatin is not only present in the vertebrate class but has also been discovered in a protozoan[7]. During evolution the structure of the 14 amino acid residue somatostatin has been surprisingly well conserved. In addition to somatostatin-14 a somatostatin molecule with 28 amino acids has been discovered in animals and also in the human which contains SS-14 at its carboxy terminal[124, 126].

Furthermore, a somatostatin-28 with the first 12 N-terminal amino acids of the N-terminus of SS-28 (SS-28 1–12) has been demonstrated in the pancreas and in the brain. If all 3 gene products are derived from the same gene and the same preprosomatostatin, which is processed differently in different cells, is not clear. It is of interest that even within the same species, more than one gene coding for preprosomatostatin has been discovered[150, 155].

6.1. Distribution of Somatostatin

Somatostatin was originally believed to be a hypothalamic peptide, which was secreted into the portal system and inhibited GH-secretion. However, it was found to be widely distributed in tissue throughout the body, particularly in the brain[116], in the peripheral nervous system, the gastrointestinal tract, and in other organs like thyroid and kidney.

Somatostatin therefore is not only a neuroendocrine hormone but also a neurotransmitter, a paracrine and an autocrine hormone. It may also act after being secreted into the lumen of the gastrointestinal tract ("lumone").

Somatostatin cells and nerve fibers are found in the hypothalamus terminating in the median eminence in agreement with the hypophysiotropic role of this hormone[116]. Cell bodies for these neurons are located in the anterior periventricular and also in the medial part of the paraventricular area of the hypothalamus. It is of interest that somatostatin containing nerve endings are also found in the suprachiasmatic nucleus which is believed to be the main area where biological rhythms are controlled[166]. Somatostatin containing cells are also found in the cortex and in the spinal cord.

In peripheral nerves somatostatin is colocalized with other peptides like endorphins and catecholamines[116, 195].

In the gastrointestinal tract somatostatin is found in the D-cells of the pancreatic islets where it controls insulin and glucagon release in a paracrine fashion[110, 122, 126]. Furthermore, somatostatin is found in the stomach and also in the intestines with decreasing tissue concentrations towards the colon[126]. Somatostatin is secreted into the interstitial space of the gut wall (paracrine secretion) but also in the lumen where it may modify acid and gastrin secretion[25].

Furthermore, somatostatin has been found in C-cells of the human thyroid, which contain also calcitonin, and in medullary thyroid carcinoma cells[126]. Furthermore it has been demonstrated in the collecting tubules of the kidney, where it may interfere with the action of antidiuretic hormone[129].

6.2. Physiological Role of Somatostatin

Somatostatin is a very potent inhibitor of hormone secretion and hormone action. In this respect it has been called "endocrine cyanide" or panhibin[94]. Though we know little about the neurotransmitter role of somatostatin, it is well established, that somatostatin is an important regulator of exocrine and endocrine pancreatic function, both being inhibited effectively[126]. It also inhibits gastrin secretion and gastrin induced hydrochloric acid secretion[25]. It inhibits gastric motility, gallbladder contraction and splanchnic blood flow. It inhibits the adsorption of nutrients from the gut and may therefore be an important regulator of food intake[145, 146].

6.2.1. Somatostatin as a Hypophysiotropic Hormone

Somatostatin inhibits GH secretion at the somatotroph level where receptors for this hormone have been demonstrated[77]. The events which occur after somatostatin has been bound to the receptor are not completely clear, though it has been shown that cAMP levels decrease in cells, which are exposed to somatostatin[126]. The pulsatile secretion of GH is not caused by pulsatile secretion of somatostatin but by the pulsatile activity of the GRH-neurons. Somatostatin seems to fluctuate in a slow pattern, modifying the amplitude of the GH-pulses induced by episodic secretion of GRH[2]. Recently, it has been shown that insulin hypoglycemia following a GRH-pulse in humans leads to a regular GH-increase[152], whereas a second GRH-pulse leads only to a blunted GH-response. This would suggest, that the hypoglycemia induced GH-response is not mediated by GRH but possibly by suppression of somatostatin release from the hypothalamus.

Somatostatin receptors have also been demonstrated on thyrotroph membranes. Accordingly, somatostatin also inhibits basal and TRH-stimulated TSH-secretion[167]. The physiological role of hypothalamic somatostatin for the regulation of TSH-secretion is not clear, particularly since a second hypothalamic hormone, dopamine, also inhibits TSH secretion[140] (Fig. 1).

6.2.2. Measurement of Somatostatin

Somatostatin, which has a half time of plasma disappearance of 2 minutes can be measured by specific, sensitive radioimmunoassays. Differences in reported concentrations may be due to varying specificity of the antibodies which may measure only somatostatin 1–14 or also somatostatin 1–28, both present in the peripheral circulation. Somatostatin

1–14 levels range between 2 and 10 pg/ml and do not reflect hypothalamic but gastrointestinal somatostatin secretion. Thus, somatostatin levels for example increase during protein rich meals[126]. Plasma measurements of somatostatin for clinical purposes are only indicated in cases in whom an excess somatostatin secretion (somatostatinoma syndrome) is suspected.

Somatostatin can also be measured in the cerebrospinal fluid[115]. It seems that in neurodegenerative diseases particularly low somatostatin-concentrations are detected[126, 136].

6.3. Pathophysiology of Somatostatin

Theoretically, acromegaly could be caused by a lack of hypothalamic somatostatin activity. However, there is no evidence that there is a hypothalamic defect of somatostatin release leading to acromegaly[96]. Furthermore, there are no endocrine data which would support that growth hormone deficiency is caused by an excessive release of somatostatin into the portal circulation. Patients with somatostatinomas—rare, malignant, endocrine active pancreatic tumours—have symptoms which are explained by somatostatin induced exocrine and endocrine pancreatic insufficiency—diarrhea, fatty stools, diabetes mellitus—and gallstones—the latter due to the hypomotility of the gallbladder due to excessive somatostatin bioactivity. The diagnosis is made by measuring excessive concentrations of circulating somatostatin in the peripheral circulation[67].

Since low concentrations of somatostatin immunoreactivity were found in the cerebrospinal fluid of patients with Alzheimer disease[136], it was speculated that presenile dementia may be due to lack of somatostatinergic neurotransmission analogous to dopamine deficiency in Parkinson's disease. Extensive research is going on extending also to psychiatric disorders[124]. Thus, in patients with depression significantly reduced concentrations of immunoreactive somatostatin were detected in the cerebrospinal fluid, which normalized after remission[126].

6.4. Clinical Use of Somatostatin

There is no evidence that somatostatin has any diagnostic value as a test substance. As mentioned, the direct measurement of somatostatin plasma levels is of minor clinical importance. Only in patients with unclassified, endocrine-active pancreatic tumours, the rare somatostatinoma syndrome can be excluded by assaying somatostatin levels[67]. However, SS 1–14 is of therapeutical value in patients with severe acute bleeding from the gastrointestinal tract due to gastroduodenal ulcers or diffuse, hemorrhagic gastrointestinal inflammation[124].

Furthermore, though this is still debated, somatostatin may be useful

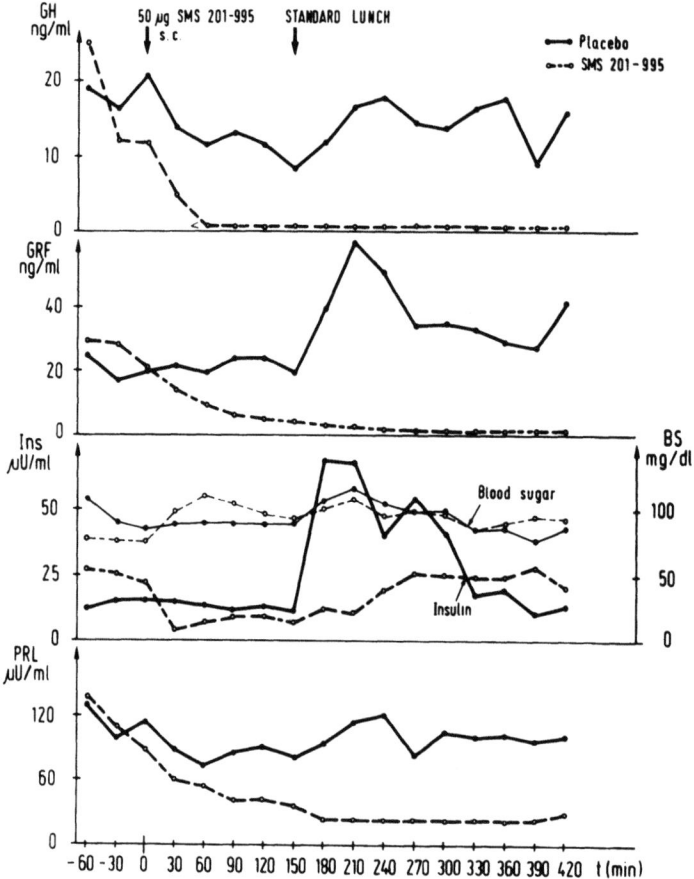

Fig. 17. GH, GRH, insulin (Ins), and prolactin (PRL) levels after sc injection of 50 µg of a somatostatin analogue (SMS 201–995) in a patient with a gut GRF-oma and liver metastases. The somatostatin analogue was injected at 9.00 a.m. and hormone levels were monitored from 8 a.m. to 5 p.m. (●——●) and compared with those of a control day (O-----O). All 4 hormones were significantly suppressed. The 15-year-old girl, who presented clinically with gigantism and suprasellar extending pituitary hyperplasia, has now been treated chronically for two years with 50 µg SMS 201–995 b.i.d. with continuous suppression of GH and GRH. (With permission from von Werder et al., 1984)

after surgery for pancreatitis and as an adjuvant therapy for postoperative pancreatic and duodenal fistulas[172].

It has also been used as an inhibitor of the exocrine pancreatic function before pancreatic transplantation for diabetes mellitus. Besides its utilization in gastrointestinal diseases, somatostatin has been used for improving blood sugar control in insulin dependent diabetics through suppression of hyperglucagonemia and growth hormone levels. However, this form of treatment has not been pursued[40, 41]. One of the problems in treating patients with somatostatin is the fact that it must be infused in a dosage of 3.5 µg per kg body weight per hour because of its short half life of 1 to 3 minutes[125, 126]. As soon as the infusion is stopped not only the somatostatin effect disappears but a rebound in the suppressed hormone secretion is observed. Therefore the development of long acting somatostatin analogues which can be given subcutaneously is of great advantage. The so-called minisomatostatin (SMS 201–995, Sandoz AG, Basle, Switzerland), an octapeptide-analogue[4], inhibits basal and GRH stimulated GH secretion[86] and gastrointestinal hormone secretion after s.c. administration in a dosage of 50 or 100 µg over a period of 6 to 8 hours[86, 121]. Long-term treatment with this compound has been shown to normalize GH-levels in active acromegaly[23, 73]. Furthermore, since the octapeptide, like the tetradecapeptide, is effective in suppressing tumorous gastrointestinal hormone secretion, it has been proven to be useful in treating surgically incurable endocrine-active gut tumours. Thus, one patient with liver metastases due to a duodenal GRH-oma and GH hypersecretion has been treated for 18 months with this octapeptide, leading to complete suppression of GRH and GH levels[189] (Fig. 17).

Native somatostatin 1–14 is also an effective analgesic substance when given intrathecally. The analgesic effect cannot be antagonized by naloxone, though the efficacy in reducing pain perception is comparable to opiates. The possible use of somatostatin in anaesthesia is under investigation[124].

7. Hypothalamic Hormones Regulating Prolactin Secretion

Prolactin is the only anterior pituitary hormone which is under predominantly inhibitory control of the hypothalamus[36, 125]. Disruption of the hypothalamic-pituitary connection or destruction of hypothalamic hypophysiotropic centres usually leads to hypothalamic or secondary anterior pituitary failure in respect to gonadotropin, GH, ACTH, and TSH secretion but to an excessive secretion of prolactin, demonstrating the importance of the hypothalamic inhibitory activity exerted on the lactotrophs. There is considerable evidence that in addition to the prolactin inhibiting factor there is a physiological prolactin releasing factor. As has been pointed out TRH stimulates PRL-secretion but may not be the physiological prolactin releasing hormone, because of the frequent dissociation between TSH and PRL-secretion.

7.1. Prolactin Inhibiting Factor

As soon as prolactin was demonstrated to be a separate anterior pituitary hormone in humans more than 15 years ago, it was also shown that dopamine and dopamine agonists inhibit human PRL secretion as has been documented earlier already in animals. Furthermore, it was shown that a number of neurons in the pituitary stalk area contain dopamine[6, 125]. These tuberoinfundibular neurons (TIDA) have close association to the portal capillaries and it was therefore not surprising that dopamine could be demonstrated in the portal blood[6]. This, and the fact that lactotroph cells have dopamine receptors and that dopamine inhibits PRL secretion is sufficient evidence for dopamine being the main hypothalamic hormone by which prolactin inhibition is mediated. It would be justified to use the name prolactin inhibiting hormone (PIH) though throughout this review the more accustomed name PIF is still used. Recently, a peptidergic prolactin inhibiting factor has been discovered, represented by an amino acid sequence which is found in the C-terminal part of the prepro-GnRH (GnRH associated peptide = GAP[107]). The role of this peptidergic hormone is open, though its close association to GnRH is intriguing.

Dopamine not only inhibits PRL-secretion but also TSH-secretion which is under predominantly stimulatory control through hypothalamic TRH but double inhibitory control via hypothalamic somatostatin and dopamine (Fig. 1). In addition, hypothalamic dopamine attenuates the frequency and the pulse amplitude of endogenous GnRH release. Elevated prolactin levels exert an inhibitory feedback effect on prolactin release by increasing hypothalamic dopamine turnover[36]. The lack of GnRH pulsatility leading to hyperprolactinemic hypogonadism therefore may be explained by elevated PRL levels leading to elevated hypothalamic dopamine concentrations which influence GnRH-secretion in the described fashion[187]. That hypothalamic dopamine in this situation cannot reach the lactotrophic adenoma may be due to local vascular changes, which according to Weiner *et al.* may be the cause of beginning prolactinoma formation[182].

7.2. Prolactin Releasing Factor (PRF)

Though endogenous TRH seems to stimulate PRL-secretion in pathological situations like primary hypothyroidism, it is probably not the physiological PRF. The most likely candidate is vasoactive intestinal peptide (VIP) which is found in the hypothalamus as well as in the portal circulation[1, 125]. Exogenously administered VIP stimulates PRL secretion with the same kinetic as other releasing hormones stimulate anterior pituitary hormone secretion[1, 80]. Furthermore it could be demonstrated that serotonin induced PRL secretion is mediated by VIP. Thus rats show a prompt rise of PRL after administration of 5-hydroxytryptamin, which can be completely abolished after passive administration with VIP antisera[156]. Though this may suggest that VIP is a physiological regulator of PRL secretion, it is certainly not proven yet[125].

7.3. Clinical Utilization of Hypothalamic Hormones Regulating PRL Secretion

It has become clear, that neither stimulation nor suppression-tests for PRL-secretion are useful for the diagnostic of disturbances of PRL secretion[187]. This pertains particularly for the differential diagnosis between autonomous PRL secretion due to a microprolactinoma with a radiologically normal sella turcica and so called functional hyperprolactinemia. However, dopamine agonists, which are derivatives of ergot alkaloids and essentially superactive PIF analogues, have become very important therapeutical tools for the management of hyperprolactinemia[187]. Thus, with these drugs the only definitely effective medical treatment of anterior pituitary hormone excess has become possible. The latter is a true alternative to the operative treatment, since it has been shown that dopamine agonists not only normalize PRL secretion but can also reduce the size of the PRL secreting adenoma[187].

Furthermore, these compounds have been used in other states of dopamine deficiency, particularly in Parkinson's disease, where they may lead to improvement of symptoms, when L-dopa, the dopamine precursor, has lost its efficacy.

8. Summary

The hypophysiotropic hormones of the hypothalamus represent a biochemical link between the central nervous system and the peripheral endocrine glands. The structure of these neurohormones has been elucidated and the genes coding the precursor hormones of the neuropeptides with the exception of TRH have been cloned. The only exception is the prolactin inhibiting factor which is no peptide but dopamine though evidence for a peptidergic PIF has been recently demonstrated.

The major hypothalamic influence on anterior pituitary cells is of stimulatory nature with the exception of the lactotrophs which are under tonic dopaminergic inhibitory hypothalamic control. In contrast to the species specificity of the biological activity of anterior pituitary hormones, there is no such species specificity in respect to bioactivity of the hypothalamic neurohormones[60]. However, changes of the primary structure of the long-chain neuropeptides CRH and GRH occur in parts of the molecule not involved with bioactivity in different species. TRH, GnRH, CRH and GRH are used clinically for the evaluation of anterior pituitary function. They can be given together in a combined pituitary function test[55,142], allowing to investigate all anterior pituitary hormones at the same time (Fig. 18). The inhibitory factors somatostatin and dopaminergic compounds do not play a diagnostic role. However, somatostatin and dopamine agonists are used for therapy of several diseases and GnRH and GRH are used for substitutional therapy. If TRH and CRH also have a therapeutical role is still undecided.

After discovery of the hypothalamic hormones it became soon clear that these hormones do not only play a hypophysiotropic role. The same neuropeptides are found in other parts of the brain as well as in the

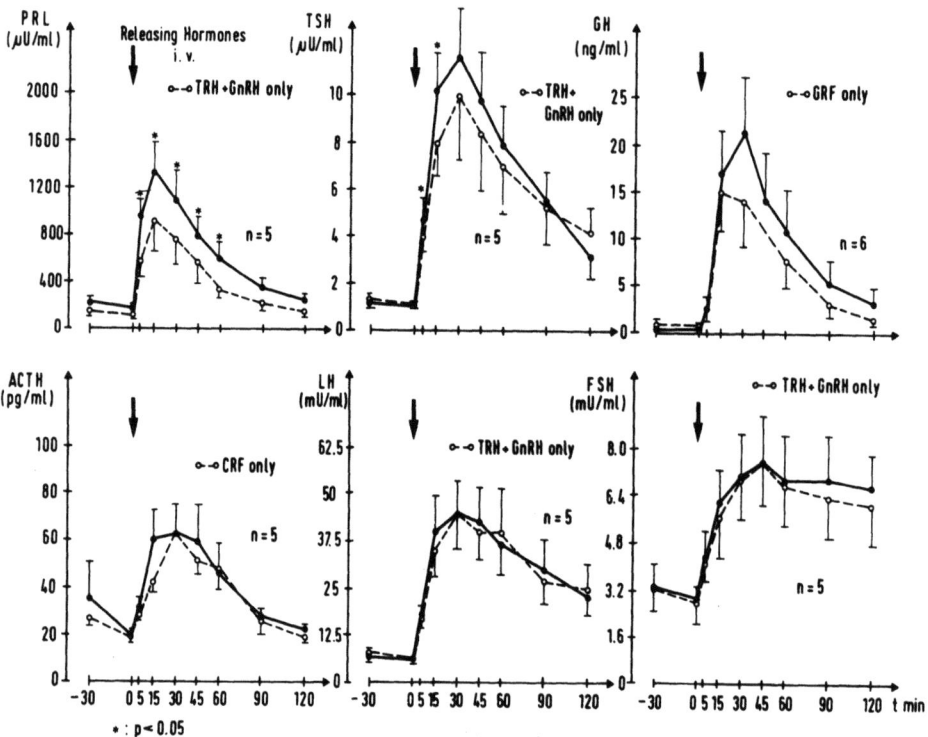

Fig. 18. Combined pituitary function test: PRL-, TSH-, GH-, ACTH-, FSH-, and LH-levels (x ± SE) after combined stimulation with 200 µg TRH, 100 µg GnRH, 100 µg CRH, and 100 µg GRH (●——●) and after separate stimulation with the appropriate releasing hormone (O-----O) in normal subjects. (With permission from Schopohl et al., 1986)

peripheral nervous system and in other peripheral organs where they play a role as neurotransmitters and modulators of peripheral organ function. With the exception of somatostatin whose gastrointestinal role is well established, we know little about the physiological role of the hypothalamic neurohormones in extrahypothalamic sites of the CNS and the periphery.

Defects in the secretion of the hypophysiotropic hormones into the portal system lead to typical clinical entities. Thus, growth hormone deficiency leading to pituitary dwarfism and hypogonadotropic hypogonadism are frequently caused by hypothalamic GRH or GnRH defici-

Table 2. *Therapy of Hypothalamic Hormone Deficiency*

Therapy	Adrenal insufficiency	Hypothyroidism	Hypogonadism	Hyposomatotropism	Diabetes insipidus
Hypothalamic neuropeptides	0	0	GnRH (pulsatile)	GRH	ADH-analogues (DDAVP)
Pituitary	0	0	hMG/hCG	hGH	carbamazepin (stimulates ADH-release)
Target gland hormones	hydrocortisone (25–30 mg/d)	thyroid hormones (T4/T3)	androgens, estrogens, progestagens	0	

Table 3. *Hypothalamic Causes of Hormone Hypersecretion*

Hypothalamic dysfunction	Early activation of the GnRH-pulse generator	GRH-hypersecretion	CRH-hypersecretion	Dopamine deficiency
Hypersecreted hormone	LH/FSH	GH	ACTH	PRL
Clinical state	precocious puberty	acromegaly, gigantism	Cushing's disease	hypogonadism
Therapy	GnRH-analogues	tumour removal, minisomatostatin DA-agonists	tumour removal, adrenolytic therapy (o,p-DDD) bilateral adrenalectomy	DA-agonists

ency. In this case the hormone deficiency can be effectively treated by substitution with the releasing hormones (Table 2). Adrenal insufficiency and hypothyroidism are less frequently caused by releasing hormone deficiency. In both situations target hormone replacement is therapy of choice regardless of the cause of glandular hypofunction (Table 2). How often hyperprolactinemia is due to hypothalamic dopamine deficiency is not clear. Regardless of its origin, dopamine agonist would be the treatment of choice in most instances.

Anterior pituitary hypersecretion may also be caused by an inappropriately enhanced secretion of releasing hormones.

Early activation of the GnRH pulse generator may cause precocious puberty. Similarly acromegaly as well as Cushing's disease may be due to enhanced secretion of GRH or CRH though the latter cannot be proven (Table 3). Only in ectopic releasing hormone hypersecretion, which has been described for GRH and CRH, releasing hormone dependent hyperplasia of the respective anterior pituitary cells can be documented.

Analogues of releasing hormones, which in the case of GnRH lead to an opposite biological effects can be used to correct pathological hypothalamic hypophysiotropic function (Table 3).

References

1. Abe H, Engler D, Molitch ME, Bollinger-Gruber J, Reichlin S (1985) Vasoactive intestinal peptide is a physiological mediator of prolactin release in the rat. Endocrinology 116: 1383–1390
2. Arimura A, Culler MD (1985) Regulation of growth hormone secretion. In: Imura H (ed). The pituitary gland. Raven Press, New York, pp 221–259
3. Baird A, Wehrenberg WB, Böhlen P, Ling N (1985) Immunoreactive and biologically active growth hormone-releasing factor in the rat placenta. Endocrinology 117: 1508–1601
4. Bauer W, Briger U, Doepfner W, Haller R, Huguenin R, Marbach P, Petcher TJ, Pless J (1982) SMS 201–995: A very potent and selective octapeptide analogue of somatostatin with prolonged action. Life Sci 31: 1133–1144
5. Beck-Peccoz P, Amr, S, Menezes-Ferreira M, Faglia G, Weintraub BD (1985) Decreased receptor binding of biologically inactive thyrotropin in central hypothyroidism. New Engl J Med 312: 1085–1090
6. Ben-Jonathan N (1985). Dopamine: A prolactin-inhibiting hormone. Endocr Rev 6: 564–589
7. Berelowitz M, LeRoith D, von Schenk H, et al. (1982) Somatostatin like immunoreactivity and biological activity is present in Tetrahymena pyriformis, a ciliated protozoan. Endocrinology 110: 1934–1944
8. Berg D, Rjosk HK, Jänicke F, von Werder K (1983) Behandlung der hyperprolaktinämischen Amenorrhoe durch pulsatile Gabe von Gonadotropin-Releasing Hormonen. Geburtshilfe Frauenheilkd 43: 686–688

9. Berger G, Trouillas J., Bloch B, Sassolas G, Berger F, Partensky C, Chayvialle JA, Brazeau P, Claustrat B, Lesbros F, Girod C (1984) Multihormonal carcinoid tumor of the pancreas secreting growth hormone-releasing factor as a cause of acromegaly. Cancer 54: 2097–2108
10. Bernutz C, Kewenig M, Horn K, Pickardt CR (1985) Detection of thyroid disorders by use of basal thyrotropin values determined with an optimized sandwich enzyme immunoassay. Clin Chem 31: 289–292
11. Besser GM, McNeilly AS, Anderson DC, Marshall JC, Harsoulis P, Hall R, Ormston BJ, Alexander L, Collins WP (1972) Hormonal responses to synthetic luteinizing hormone and follicle stimulating hormone-releasing hormone in man. Br Med J 3: 267–271
12. Bloch B, Brazeau P, Ling N, Böhlen P, Esch F, Wehrenberg WB, Benoit R, Benoit FR, Bloom F, Guillemin R (1983) Immunohistochemical detection of growth hormone-releasing factor in brain. Nature 301: 607
13. Borges JLC, Gelato MC, Rogol AD, Vance ML, MacLeod RM, Loriaux DL, Rivier J, Blizzard RM, Furlanetto R, Evans WS, Kaiser DL, Merriam GR (1983) Effects of human pancreatic tumor growth hormone releasing factor on growth hormone and somatomedin C levels in patients with idiopathic growth hormone deficiency. Lancet 2: 119
14. Borowski GD, Garofano CD, Rose LI, Lewy RA (1984) Blood pressure response to thyrotropin-releasing hormone in euthyroid subjects. J Clin Endocrinol Metab 58: 197
15. Brazeau P, Vale W, Burgus R, Ling N, Rivier J, Guillemin R (1973) Hypothalamic polypeptide that inhibits the secretion of pituitary immunoreactive growth hormone. Science 179: 77–79
16. Brownstein MJ, Palkovits M, Saavedra JM, Bassiri R, Utiger RD (1974) Thyrotropin-releasing hormone in specific nuclei of rat brain. Science 185: 267–269
17. Burger HG, Patel YC (1977) Thyrotropin releasing hormone – TSH. Clin Endocrinol Metab 6: 83–100
18. Burgus R, Butcher M, Amoss M, Ling N, Monahan M, Rivier J, Fellows R, Blackwell R, Vale W, Guillemin R (1972) Primary structure of the hypothalamic luteinizing hormone-releasing factor (LRF) of ovine origin. Proc Natl Acad Sci USA 69: 278–282
19. Burgus R, Dunn TF, Desiderio D, Ward DN, Vale W, Guillemin R (1970) Characterization of the hypothalamic hypophysiotropic TSH-releasing factor (TRF) of ovine origin. Nature 226: 321–325
20. Carey RM, Varma SK, Drake CR, Thorner MO, Kovacs K, Rivier J, Vale W (1984) Ectopic secretion of corticotropin-releasing factor as a cause of Cushing's syndrome. N Engl J Med 311: 13–20
21. Charlton BG, Leake A, Ferrie IN, Linton EA, Lowry PJ (1986) Corticotropin-releasing factor in plasma of depressed patients and controls. Lancet 1: 161–162
22. Chapman AJ, Williams G, Hockley AD, London DR (1985) Pituitary apoplexy after combined test of anterior pituitary function. Br Med J 291: 26

23. Ch'ng LJC, Sandler LM, Kraenzlin ME, Burrin M, Joplin GF, Bloom SR (1985) Long term treatment of acromegaly with a long acting analogue of somatostatin. Br Med J 290: 284–285
24. Chrousos GP, Schulte HM, Oldfield EH, Gold PW, Cutler GB, Loriaux DL (1984) The corticotropin-releasing factor stimulation test: An aid in the evaluation of patients with Cushing's syndrome. N Engl J Med 310: 622–626
25. Colturi TJ, Unger RH, Feldman M (1984) Role of circulating somatostatin in regulation of gastric acid secretion, gastrin release, and islet cell function. J Clin Invest 74: 417–423
26. Deuben RR, Meites J (1964) Stimulation of pituitary growth hormone release by a hypothalamic extract *in vitro*. Endocrinology 74: 408
27. Dolva LO, Riddervold F, Thorsen RK (1983) Side effects of thyrotropin releasing hormone. Br Med J 287: 532
28. Duick DS, Wahner HW (1979). Thyroid axis in patients with Cushing's syndrome. Arch Int Med 139: 767–772
29. Engel WK, Siddique T, Nicoloff JT (1983) Effect on weakness and spasticity in amyotrophic lateral sclerosis of thyrotropin-releasing hormone. Lancet 1: 73–75
30. Evans LEJ, Hunter P, Hall R, Johnston M, Roy VM (1975) A double blind trial of intravenous thyrotropin-releasing hormone in the treatment of reactive depression. Br J Psychiatry 127: 227–230
31. Faden AI, Jacobs TP, Holaday JW (1981) Thyrotropin-releasing hormone improves neurologic recovery after spinal trauma in cats. N Engl J Med 305: 1063–1067
32. Faglia G, Beck-Peccoz P, Ferrari C, Ambrosi B, Spada A, Travaglini P (1973) Plasma growth hormone response to thyrotropin-releasing hormone in acromegaly. J Clin Endocrinol Metab 36: 1259–1262
33. Faglia C, Bitensky L, Pinchera A, Ferrari C, Parrachi A, Beck-Peccoz P, Ambrosi B, Spada A (1979) Thyrotropin secretion in patients with central hypothyroidism: Evidence for reduced biological activity of immunoreactive thyrotropin. J Clin Endocrinol Metab 48: 989–998
34. Fahlbusch R, Pickardt CR (1975) The effect of intraventricular TRH in patients with diseases of the hypothalamic and pituitary region. Acta Endocrinol 193: 90
35. Ferin M, Van Vugt D, Wardlaw S (1984) The hypothalamic control of the menstrual cycle and the role of endogenous opioid peptides. Rec Prog Horm Res 40: 441
36. Flückiger E, del Pozo E, von Werder K (1982). Prolactin, physiology, pharmacology and clinical findings. Monographs in Endocrinology 23, Springer, Berlin Heidelberg New York
37. Forest MG (1985) Sexual maturation of the hypothalamus: Pathophysiological aspects and clinical implications. Acta Neurochir (Wien) 75: 23–42
38. Franchimont P, Becker H, Ernould CH, Thys CH, Demoulin A, Bourguignon JP, Legros JJ, Valcke JC (1974) The effect of hypothalamic luteinizing hormone releasing hormone (LHRH) on plasma gonadotropin levels in normal subjects. Clin Endocrinol 3: 27–29
39. Furutani Y, Morimoto Y, Shibahara S, *et al.* (1983) Cloning and sequence

analysis of cDNA for ovine corticotropin-releasing factor precursor. Nature 301: 537–540
40. Gerich JE, Lorenzi M, Schneider V, Forsham PH (1974) Effect of somatostatin on plasma glucose and insulin responses to glucagon and tolbutamide in man. J Clin Endocrinol Metab 39: 1057–1060
41. Gerich JE, Lorenzi M, Bier D, Schneider V, Tsalikia E, Karam J, Forsham PH (1975) Prevention of human diabetic ketoacidosis by somatostatin. Evidence for an essential role of glucagon. N Engl J Med 292: 985–989
42. Giguère V, Labrie F, Coté J, Coy DH, Sueiras-Diaz J, Schally AV (1982) Stimulation of cyclic AMP accumulation and adrenocorticotropin release by synthetic ovine corticotropin-releasing factor in rat anterior pituitary cells: site of glucocorticoid action. Proc Natl Acad Sci USA 79: 3466
43. Gillies GE, Linton EA, Lowry PJ (1982) Corticotropin releasing activity of the new CRF is potentiated several times by vasopressin. Nature 299: 355–357
44. Gold PW, Goodwin FK, Wher T, Rebar R (1978) Pituitary thyrotropin response to thyrotropin-releasing hormone in affective illness: Relationship to spinal fluid amine metabolism. Am J Psychiatry 134: 1028–1031
45. Goodman RH, Jacobs JW, Dee PC, Habener JF (1982) Somatostatin-28 encoded in a cloned cDNA obtained from a rat medullary thyroid carcinoma. J Biol Chem 257: 1156–1159
46. Griffiths EC, Bennett GW (1983) Thyrotropin releasing hormone. Raven Press, New York
47. Gubler U, Monahan JJ, Lomedico PT, Bhatt RS, Collier KJ, Hoffman BJ, Böhlen P, Esch F, Ling N, Zeytin F, Brazeau P, Poonian MS, Gage LP (1983) Cloning and sequence analysis of cDNA for the precursor of human growth hormone-releasing factor, somtocrinin. Proc Natl Acad Sci USA 80: 4311
48. Guillemin R (1978) The brain as an endocrine organ. Neurosci Res Program Bull 16: [Suppl] 1–25
49. Guillemin R, Brazeau P, Böhlen P, Esch F, Ling N, Wehrenberg WB (1982) Growth hormone releasing factor from a human pancreatic tumor that caused acromegaly. Science 218: 585–587
50. Guillemin R, Brazeau P, Böhlen P, Esch F, Ling N, Wehrenberg WB, Bloch B, Mougin C, Zeytin F, Baird A (1984) Somatocrinin, the growth hormone releasing factor. Recent Prog Horm Res 40: 233–286
51. Habermann J, Leisner B, Witte A, Pickardt CR, Scriba PC (1982) Iodine contamination as a cause of hyperthyroidism or lack of TSH response to TRH stimulation (Results based on a screening investigation). J Endocrinol Invest 5: 153–156
52. Harris GW (1948) Neural control of the pituitary gland. Physiol Rev 28: 134–179
53. Hinkle PM, Woroch EL, Tashjian AH jr (1974) Receptor-binding affinities and biological activities of analogs of thyrotropin-releasing hormone in prolactin-producing pituitary cells in culture. J Biol Chem 249: 3085–3090
54. Hofmann AR, Crowley WF (1982) Induction of puberty in men by longterm pulsatile administration of low-dose gonadotropin-releasing hormone. N Engl J Med 307: 1237–1241

55. Holl R, Fehm HL, Hetzel WD, Heinze E, Voigt KH (1985) Globaler Hypophysenstimulationstest mit Releasing-Hormonen. Dtsch Med Wochenschr 110: 953–955
56. Horn K, Erhardt F, Fahlbusch R, Pickardt CR, von Werder K, Scriba PC (1976) Recurrent goiter, hyperthyroidism, galactorrhea and amenorrhea due to a thyrotropin and prolactin-producing pituitary tumour. J Clin Endocrinol Metab 43: 137–143
57. Imaki I, Shibasaki T, Shizume K, Masuda A, Hotta M, Kiyosawa Y, Jibiki K, Demura H, Tsushima T, Ling N (1985) The effect of free fatty acids on growth hormone (GH)-releasing hormone-mediated GH secretion in man. J Clin Endocrinol Metab 60: 290–293
58. Jackson IMD (1982) Thyrotropin releasing hormone. N Engl J Med 306: 145–155
59. Jackson IMD (1980) Diagnostic tests for the evaluation of pituitary tumors. In: Post KD, Jackson IMD, Reichlin S (eds). The pituitary adenoma. Plenum Press, New York
60. Jackson IMD (1978) Phylogenetic distribution and function of the hypophysiotropic hormones of the hypothalamus. Am Zoologist 18: 385–399
61. Jackson IMD, Reichlin S (1974) Thyrotropin releasing hormone (TRH): Distribution in the brain, blood and urine of the rat. Life Sci 14: 2253–2266
62. Jacobs LS, Snyder PJ, Utiger RD, Daughaday WH (1973) Prolactin response to thyrotropin releasing hormone in normal subjects. J Clin Endocrinol Metab 36: 1069–1073
63. Kastin AJ, Ehrensing RH, Schalch DS, Anderson MS (1972) Improvement in mental depression with decreased thyrotropin response after administration of thyrotropin-releasing hormone. Lancet 2: 740–742
64. Knobil E (1980) The neuroendocrine control of the menstrual cycle. Recent Prog Horm Res 36: 53–88
65. Knobil E, Plant TM, Wildt L, Blecher PE, Marshall G (1980) Control of the rhesus monkey menstrual cycle: permissive role of the hypothalamic gonadotrophin-releasing hormone. Science 207: 1371
66. Kourides IA, Ridgway EC, Weintraub BC, Bigos ST, Gershengorn MC, Maloof F (1977) Thyrotropin-induced hyperthyroidism: Use of alpha and beta subunit levels to identify patients with pituitary tumours. J Clin Endocrinol Metab 45: 534–543
67. Krejs GJ, Orci L, Conlon JM, Ravazzola M, Davis GR, Raskin P, Collins SM, McCarthy DM, Baetens D, Rubenstein A, Aldor AM, Unger RH (1979) Somatostatinoma syndrome. Biochemical, morphological and clinical features. N Engl J Med 301: 285–292
68. Krieger DT, Condon EM (1978) Cyproheptadine treatment of Nelson's syndrome: restoration of plasma ACTH circadian periodicity and reversal of response of TRF. J Clin Endocrinol Metab 46: 349–352
69. Krieger DT (1983) Physiopathology of Cushing's disease. Endocr Rev 4: 22
70. Krulich L, Dhariwal APS, McCann SM (1968) Stimulatory and inhibitory effects of purified hypothalamic extracts on growth hormone release from rat pituitary *in vitro*. Endocrinology 83: 783–790
71. Labrie F, Dupont A, Belanger A (1984) Spectacular response to combined

antihormonal treatment in advanced prostate cancer. In: Labrie F, Proulx L (eds) Endocrinology. Excerpta Medica, Amsterdam New York Oxford, pp 450–453
72. Labrie F, Veilleux R, Lefèvre G, Coy DH, Sueiras-Diaz J, Schally AV (1982) Corticotropin-releasing factor stimulates accumulation of adenosine 3',5'-monophosphate in rat pituitary corticotrophs. Science 216: 1007
73. Lamberts SWJ, Uitterlinden P, Verschoor L, van Dongen KJ, Del Pozo E (1985) Long-term treatment of acromegaly with the somatostatin analogue SMS 201-995. N Engl J Med 313: 1576–1580
74. Lamberts SWJ, Verleun T, Oosterom R, DeJong F, Hackens WHL (1984) Corticotropin-releasing factor (ovine) and vasopressin exert a synergistic effect on adrenocorticotropin release in man. J Clin Endocrinol Metab 58: 298–303
75. Lance VA, Murphy WA, Sueiras-Diaz J, Coy EH (1984) Superactive analogs of growth hormone releasing factor (1–29)-amide. Biochem Biophys Res Commun 119: 265
76. Laron Z, Keret R, Bauman B, Pertzelan A, Ben-Zeev Z, Olsen DB, Comaru-Schally AM, Schally AV (1984) Differential diagnosis between hypothalamic and pituitary hGH deficiency with the aid of synthetic GH-RH 1–44. Clin Endocrinol 21: 9
77. Leitner JW, Rifkin RM, Maman A, Sussman KE (1979) Somatostatin binding to pituitary plasma membranes. Biochem Biophys Res Commun 87: 919–927
78. Leyendecker G, Wildt L (1983) Induction of ovulation with chronic-intermittent (pulsatile) administration of GnRH in hypothalamic amenorrhea. J Reprod Fertil 69: 397
79. Leyendecker G, Wildt L, Hansmann M (1980) Pregnancies following chronic intermittent (pulsatile) administration of GnRH by means of portable pump (Zyklomat)—a new approach to the treatment of infertility in hypothalamic amenorrhea. J Clin Endocrinol Metab 51: 1214–1216
80. Lightman S, Unwin RJ, Graham K, Dimaline R, Mc Garrick G (1984) Vasoactive intestinal polypeptide stimulation of prolactin release and renin activity in normal man and patients with hyperprolactinemia. Effects of pretreatment with bromocriptine and dexamethasone. Eur J Clin Invest 14: 444–448
81. Linde R, Doelle GC, Alexander N, Kirchner F, Vale W, Rivier J, Rabin D (1981) Reversible inhibition of testicular steroidogenesis and spermatogenesis by a potent gonadotropin-releasing hormone agonist in normal men. N Engl J Med 305: 663–667
82. Ling N, Esch F, Böhlen P, Brazeau P, Wehrenberg WB, Guillemin R (1984) Isolation, primary structure, and synthesis of human hypothalamic somatocrinin: growth hormone-releasing factor. Proc Natl Acad Sci USA 81: 4302–4306
83. Ling N, Zeytin F, Böhlen P, Esch F, Brazeau P, Wehrenberg WB, Baird A, Guillemin R (1985) Growth hormone releasing factor. Ann Rev Biochem 54: 403–423
84. Liuzzi A, Chiodini PG, Botalla L, Silvestrini F, Müller EE (1974) Growth

hormone (GH)-releasing activity of TRH and GH lowering effect of dopaminergic drugs in acromegaly: homogeneity in the two responses. J Clin Endocrinol Metab 39: 871–874
85. Losa M, Bock L, Schopohl J, Stalla GK, Müller OA, von Werder K (1984) Growth hormone releasing factor infusion does not sustain elevated GH-levels in normal subjects. Acta Endocrinol 107: 462–470
86. Losa M, Müller OA, Sobieszczyk, von Werder K (1985) Interaction between growth hormone releasing factor (GRF) and somatostatin analogue (SMS 201–995) in normal subjects. Clin Endocrinol 23: 715–720
87. Losa M, Schopohl J, Stalla GK, Müller OA, von Werder K (1985) Growth hormone releasing factor-test in acromegaly: Comparison with other dynamic tests. Clin Endocrinol 23: 99–109
88. Losa M, Schopohl J, Müller OA, von Werder K (1984) Stimulation of growth hormone secretion with human growth hormone releasing factors (GRF 1–44, GRF 1–40, GRF 1–29) in normal subjects. Klin Wochenschr 62: 1140–1143
89. Losa M, Stalla GK, Müller OA, von Werder K (1983) Human pancreatic growth hormone releasing factor (hpGRF): Dose response of GRF- and GH-levels. Klin Wochenschr 61: 1249–1253
90. Mansfield MJ, Beardsworth DE, Loughlin JS, Crawford JD, Bode HH, Rivier J, Vale W, Kushner DV, Crigler JF, Crowley WF (1983) Long-term treatment of central precocious puberty with longacting analogue of luteinizing hormone-releasing hormone. N Engl J Med 109: 1286–1290
91. Masuda A, Shibasaki T, Nakahara M, Imahi T, Kiyosawa Y, Jikibi K, Demura H, Shizume K, Ling N (1985) The effect of glucose on growth hormone (GH)-releasing hormone-mediated GH-secretion in man. J Clin Endocrinol Metab 60: 523–526
92. Matsuo H, Baba Y, Nair RMG, Arimura A, Schally AV (1971) Structure of the porcine LH- and FSH-releasing hormone. I. Proposed amino acid sequence. Biochem Biophys Res Commun 43: 1374–1439
93. McCann SM (1983) Progress in neuroendocrinology: LH releasing (LHRH), basic and clinical aspects. J Endocrinol Invest 6: 243–251
94. McCann SM, Krulich L, Negro-Vilar A, Ojeda SR, Vijayan E (1980) Regulation and function of panhibin (somatostatin). Adv Biochem Psychopharmacol 22: 131–143
95. McCann SM (1971) Mechanism of action of hypothalamic hypophyseal stimulating and inhibiting hormones. In: Ganong WF, Martini L (eds) Frontiers in neuroendocrinology. Oxford University Press, New York, p 209
96. Melmed S, Braunstein GD, Horvath E, Ezrin C, Kovacs K (1983) Pathophysiology of acromegaly. Endocr Rev 4: 271
97. Melmed S, Ezrin C, Kovacs K, Goodman RS, Frohman LA (1985) Acromegaly due to secretion of growth hormone by an ectopic pancreatic islet-cell tumor. N Engl J Med 312: 9
98. Metcalf G, Dettmar PW (1981) Is thyrotropin-releasing hormone an endogenous ergotropic substance in the brain? Lancet 1: 586–589
99. Morley JE (1981) Neuroendocrine control of thyrotropin secretion. Endocr Rev 2: 396

100. Morris DV, Adeniyi-Jones R, Wheeler M, Sonksen P, Jacobs HS (1984) The treatment of hypogonadotrophic hypogonadism in men by the pulsatile infusion of luteinizing hormone-releasing hormone. Clin Endocrinol 21: 189–200
101. Moss RL, McCann SM (1973) Induction of mating behavior in rats by luteinizing hormone releasing factor. Science 181: 177
102. Müller OA, Dörr HG, Hagen B, Stalla GK, von Werder K (1982) Corticotropin releasing factor (CRF)-stimulation test in normal controls and patients with disturbances of the hypothalamo pituitary-adrenal axis. Klin Wochenschr 60: 1485–1491
103. Müller OA, Stalla GK, Hartwimmer J, Schopohl J, von Werder K (1985) Corticotropin releasing factor (CRF): Diagnostic implications. Acta Neurochir (Wien) 75: 49–59
104. Müller OA, Stalla GK, von Werder K (1983) Corticotropin releasing factor: A new tool for the differential diagnosis of Cushing's syndrome. J Clin Endocrinol Metab 57: 227–229
105. Nahagawa K, Akikawa K, Matsubara M, Kubo M (1985) Effect of dexamethasone on growth hormone (GH) response to growth hormone releasing hormone in acromegaly. J Clin Endocrinol Metab 60: 306–310
106. Naor Z, Catt KJ (1980) Independent actions of gonadotropin releasing hormone upon cyclic GMP production and LH release. J Biochem 225: 342
107. Nikolics K, Mason AJ, Szönyi E, Ramachandran J, Seeburg PH (1985) A prolactin-inhibiting factor within the precursor for human gonadotropin-releasing hormone. Nature 316: 511
108. Nillius SJ, Bergquist C, Wide L (1978) Inhibition of ovulation in women by chronic treatment with a stimulatory LRH analogue—a new approach to birth control? Contraception 17: 537–544
109. Odell WD, Swerdloff RS (1978) Abnormalities of gonadal function in men. Clin Endocrinol 8: 149–180
110. Orci L, Baetens D, Dubois MP, Rufener C (1975) Evidence for the D-cell of the pancreas secreting somatostatin. Horm Metab Res 7: 400–402
111. Orth DN (1984) The old and the new in Cushing's syndrome. N Engl J Med 310: 649–651
112. Orth DN, Jackson RV, DeCherney GS, DeBold CR, Alexander AN, Island DP, Rivier J (1983) Effect of synthetic ovine corticotropin-releasing factor. Dose response of plasma adrenocorticotropin and cortisol. J Clin Invest 71: 587–595
113. Page RB (1982) Pituitary blood flow. Am J Physiol 243 (Endocrinol Metab 6) E427–E442
114. Pandol SJ, Seifert H, Thomas MW, Rivier J, Vale W (1984) Growth hormone releasing factor stimulates pancreatic enzym secretion. Science 225: 326–328
115. Patel YC, Rao K, Reichlin S (1977) Somatostatin in human cerebrospinal fluid. N Engl J Med 296: 529–533
116. Pelletier G (1980) Immunohistochemical localization of somatostatin. Prog Histochem Cytochem 12 (3): 1–41

117. Penny ES, Penman E, Price J, Rees LH, Sopwith AM, Wass JAH, Lytras N, Besser GM (1984) Circulating growth hormone releasing factor concentrations in normal subjects and patients with acromegaly. Br Med J 289: 453
118. Philips HS, Hostetter G, Kerdelhue B, Kozlowski GB (1980) Immunocytochemical localization of LHRH in central olfactory pathways of hamster. Brain Res 193: 574
119. Pickardt CR, Scriba PC. TRH (1985) Pathophysiologic and clinical implications. Acta Neurochir (Wien) 75: 43–48
120. Pieters GFFM, Hermus ARMM, Smals AGH, Bertelink AKM, Benraad THJ, Kloppenborg PWC (1983) Responsiveness of the hypophyseal-adrenocortical axis to corticotropin releasing factor in pituitary-dependent Cushing's disease. J Clin Endocrinol Metab 57: 513–516
121. Plewe G, Beyer J, Krause U, Neufeld M, Del Pozo E (1984) Long acting and selective suppression of growth hormone secretion by somatostatin analogue SMS 201-995 in acromegaly. Lancet 2: 782–784
122. Polak JM, Pearse AGE, Grimelius L, Bloom SR, Arimura A (1975) Growth-hormone release-inhibiting hormone in gastrointestinal and pancreatic D cells. Lancet 5: 1220–1222
123. Proulx L, Giguere V, Coté J, Labrie F (1984) Multiple factors involved in the control of ACTH secretion. J Endocrinol Invest 7: 257–263
124. Raptis S, Rosenthal J, Gerich JE (eds) (1974) Proceedings of the 2nd International Symposium on somatostatin. Attempto, Tübingen University Press, Tübingen, FRG
125. Reichlin S (1985) Neuroendocrinology, In: Williams textbook of endocrinology, 7th ed, Saunders WB, Philadelphia, pp 492–567
126. Reichlin S (1983) Somatostatin. N Engl J Med 309: 1495–1501, 1556–1563
127. Reichlin S (1960). Growth and the hypothalamus. Endocrinology 67: 760
128. Reichlin S, Martin JB, Jackson IMD (1978). Regulation of thyroid-stimulating hormone (TSH) secretion. In: Jeffcoate SL, Hutchinson JSM (eds). The endocrine hypothalamus. Academic Press, New York
129. Reid IA, Rose JC (1977) An intrarenal effect of somatostatin on water excretion. Endocrinology 100: 782–785
130. Richter D, Ivell R (1985) Gene organization, biosynthesis, and chemistry of neurohypophyseal hormones. In: Imura H (ed). The pituitary gland. Raven Press, New York, pp 127–148
131. Rivier J, Rivier C, Vale W (1984) Synthetic competitive antagonists of corticotropin-releasing factor: Effect on ACTH secretion in the rat. Science 224: 889–891
132. Rivier C, Vale W (1983) Interaction of corticotropin-releasing factor and arginine vasopressin on adrenocorticotropin secretion *in vivo*. Endocrinology 113: 939
133. Rjosk HK, von Werder K, Fahlbusch R (1976) Hyperprolaktinämische Amenorrhoe. Geburtshilfe Frauenheilkd 36: 575–587
134. Robert JF, Quigley ME, Yen SSC (1981) Endogenous opiates modulate pulsatile luteinizing hormone release in humans. J Clin Endocrinol Metab 52: 583–585
135. Root AW, Snyder PJ, Rezvani I, DiGeorge AM, Utiger RD (1973) Inhibition

of thyrotropin-releasing hormone-mediated secretion of thyrotropin by human growth hormone. J Clin Endocrinol Metab 36: 103–107

136. Rossor MN, Emson PC, Mountjoy CQ, Roth M, Iversen LL (1980) Reduced amounts of immunoreactive somatostatin in the temporal cortex in senile dementia of Alzheimer type. Neurosci Lett 20: 373–377

137. Saito H, Saito S, Yamazaki R, Hoso E (1984) Clinical value of radioimmunoassay of plasma growth hormone-releasing factor. Lancet 2: 401

138. Samson WK, McCann SM, Chud L, Dudley CA, Moss RL (1980) Intra and extrahypothalamic luteinizing hormone releasing hormone distribution in the rat with special reference to mesencephalic sites which contain both LHRH and single neurons responsive to LHRH. Neuroendocrinology 31: 66

139. Sandow J (1983) Clinical applications of LHRH and its analogues. Clin Endocrinol 18: 571–592

140. Scanlon MF, Weightman DR, Shale DJ, More B, Heath M, Snow MH, Lewis M, Hall R (1979) Dopamine is a physiological regulator of thyrotropin (TSH) secretion in normal man. Clin Endocrinol 10: 7–15

141. Scheithauer BW, Carpenter PC, Bloch B, Brazeau P (1984) Ectopic secretion of a growth hormone releasing factor. Am J Med 76: 605–616

142. Schopohl J, Losa M, König A, Müller OA, Stalla GK, von Werder K (1986) Combined pituitary function test with four hypohalamic releasing hormones. Klin Wochenschr 64: 314–318

143. Schürmeyer TH, Knuth UA, Freischem CW, Sandow J, Bint Akhtar F, Nieschlag E (1984) Suppression of pituitary and testicular function in normal men by constant gonadotropin-releasing hormone agonist infusion. J Clin Endocrinol Metab 59: 19–24

144. Schulte EH, Loriaux DL (1983) Safety of corticotropin-releasing factor. Lancet 1: 1222

145. Schusdziarra V, Rouillar D, Unger RH (1979) Oral administration of somatostatin reduces postprandial plasma triglycerides, gastrin and gut glucagon-like immunoreactivity. Life Sci 24: 1595–1600

146. Schusdziarra V, Zyznar E, Rouiller D, et al. (1980) Splanchnic somatostatin: a hormonal regulator of nutrient homeostasis. Science 207: 530–532

147. Selye H (1936) A syndrome produced by diverse noxious agents. Nature 138: 32

148. Sharp PS, Foley K, Chadal P, Kohner EM (1984) The effect of plasma glucose on the growth hormone response to human pancreatic growth hormone releasing factor in normal subjects. Clin Endocrinol 20: 497–501

149. Sheldon H (1985) Growth hormone and prolactin: Chemistry, gene organization, biosynthesis, and regulation of gene expression. In: Imura H (ed). The pituitary gland. Raven Press, New York, pp 57–82

150. Shen LP, Pictet RL, Rutter WJ (1982) Human somatostatin I: sequence of the cDNA. Proc Natl Acad Sci USA 79: 4575–4579

151. Shibahara S, Morimoto Y, Furutani Y, et al. (1983) Isolation and sequence analysis of the human corticotropin-releasing factor precursor gene. EMBO J 2: 775–779

152. Shibasaki T, Hotta M, Masuda A, Imaki T, Obara N, Demura H, Ling N, Shizume K (1985) Plasma GH responses to GRH and insulin induced hypoglycemia in man. J Clin Endocrinol Metab 60: 1265–1267
153. Shibasaki T, Kiyosawa Y, Masuda A, Nakahara M, Imaki T, Wakabayashi I, Demura H, Shizume K, Ling N (1984) Distribution of growth hormone releasing hormone-like immunoreactivity in human tissue extracts. J Clin Endocrinol Metab 59: 263–268
154. Shibasaki T, Shizume K, Masuda A, Nakahara M, Hizuka N, Miyakawa M, Takano K, Demura H, Wakabayashi I, Ling N (1984) Plasma growth hormone response to growth hormone releasing factor in acromegalic patients. J Clin Endocrinol Metab 58: 215
155. Shields D, Warren TG, Green RF, Roth SE, Brenner MJ (1981) The primary events in the biosynthesis and post-translational processing of different precursors to somatostatin. In: Rich DH, Cross E (eds). Peptides: Synthesis-structure-function: Proceedings of the seventh American Peptide Symposium. Pierce Chemical Company, Rockford, Ill, pp 471–479
156. Shimatsu A, Kato Y, Ohta H, Tojo K, Kabayama Y, Inoue T, Imura H. Involvement of vasoactive intestinal polypeptide in serotoninergic stimulation of prolactin secretion in rats. In: Mac Leod RM, Thorner MO, Scapagnini U (eds), Prolactin, basic and clinical correlates, Fidia Research Series, Liviana Press, Padova, pp 73–78
157. Siler-Khodr TM (1983) Hypothalamic-like peptides of the placenta. Seminars in Reprod Endocrin 1: 321–333
158. Snyder PJ, Jacobs LS, Rabello MM, Sterling FH, Shore RN, Utiger RD, Daughaday WH (1974) Diagnostic value of thyrotropin-releasing hormone in pituitary and hypothalamic diseases: Assessment of thyrotropin and prolactin secretion in 100 patients. Ann Intern Med 81: 751–757
159. Snyder PJ, Jacobs LS, Utiger RD, Daughaday WH (1973) Thyroid hormone inhibition of the prolactin response to thyrotropin-releasing hormone. J Clin Invest 52: 2324–2329
160. Snyder G, Naor Z, Fawcett CP, McCann SM (1980) Gonadotropin release and cyclic nucleotides: evidence for LHRH-induced elevation of cyclic GMP levels in gonadotrophs. Endocrinology 107: 1627
161. Snyder PJ, Utiger RD (1972) Response to thyrotropin releasing hormone (TRH) in normal man. J Clin Endocrinol Metab 34: 380–385
162. Sopwith AM, Penny ES, Besser GM, Rees LH (1985) Stimulation by food of peripheral plasma immunoreactive growth hormone releasing factor. Clin Endocrinol 22: 337–340
163. Stalla GK, Hartwimmer J, Kaliebe T, Müller OA (1985) Radioimmunoassay of human corticotropin releasing factor (hCRF). Acta Endocr (Kbh) Suppl. 108: 18–19
164. Stalla GK, Hartwimmer J, von Werder K, Müller OA (1984) Ovine (o) and human (h) corticotropin releasing factor (CRF) in man: CRF-stimulation and CRF-immunoreactivity. Acta Endocrinol 106: 289–297
165. Suda T, Tomori N, Yajima F, Sumitomo T, Nakagami Y, Ushiyama T, Demura H, Shizume K (1985) Immunoreactive corticotropin-releasing factor in human plasma. J Clin Invest 76: 2026–2029

166. Takahashi JS, Zatz M (1982) Regulation of circadian rhythmicity. Science 217: 1104–1111
167. Tanjasiri P, Kozbur X, Florsheim WH (1976) Somatostatin in the physiologic feedback control of thyrotropin secretion. Life Sci 19: 657–660
168. Thorner MO, Frohman LA, Leong DA, Thominet J, Downs T, Hellman P, Chitwood J, Vaughan JM, Vale W, Besser GM, Lytras N, Edwards CRV, Schaff M, Gelato M, Krieger DT, Marcovitz S, Ituarte E, Boyd AE, Malarkey WB, Blackard WG, Prioleau G, Melmed S, Charest NJ (1984) Extrahypothalamic growth-hormone-releasing factor (GRF) secretion is a rare cause of acromegaly: plasma GRF levels in 177 acromegalic patients. J Clin Endocrinol Metab 59: 846–849
169. Thorner MO, Perryman RL, Cronin MJ, Rogol AD, Draznin M, Johanson A, Vale W, Horvath E, Kovacs K (1982) Somatotroph hyperplasia: successful treatment of acromegaly by removal of a pancreatic islet tumor secreting a growth hormone-releasing factor. J Clin Invest 70: 965–977
170. Thorner MO, Reschke J, Chitwood J, Rogol AD, Furlanetto R, Rivier J, Vale W, Blizzard RM (1985) Acceleration of growth in two children treated with human growth hormone-releasing factor. N Engl J Med 312: 4–9
171. Thorner MO, Spiess J, Vance ML, Rogol AD, Kaiser DL, Webster JD, Rivier J, Borges JL, Bloom SR, Cronin MJ, Evans WS, Mac Leod RM, Vale W (1983) Human pancreatic growth hormone releasing factor selectively stimulates growth hormone secretion in man. Lancet 1: 24–28
172. Usadel KH, Leuschner U, Überla KK (1980) Treatment of acute pancreatitis with somatostatin: a multicenter double-blind trial. N Engl J Med 303: 999–1000
173. Vale W, Rivier C, Brown MR, Spiess J, Koob G, Swanson L, Bilezikjian L, Bloom F, Rivier J (1983) Chemical and biological characterization of corticotropin releasing factor. Recent Prog Horm Res 39: 245–270
174. Vale W, Spiess J, Rivier C, Rivier J (1981) Characterization of a 41-residue ovine hypothalamic peptide that stimulates secretion of corticotropin and β-endorphine. Science 213: 1394–1397
175. Vale W, Vaughan J, Yamamoto G, Spiess J, Rivier J (1983) Effects of synthetic human pancreatic (tumor) GH releasing factor and somatostatin, triiodothyronine and dexamethasone on GH-secretion *in vitro*. Endocrinology 112: 1553–1555
176. Van Cauter E, Refetoff S (1985) Evidence for two subtypes of Cushing's disease based on the analysis of episodic cortisol secretion. N Engl J Med 312: 1343–1349
177. Vance ML, Kaiser DL, Evans WS, Furlanetto R, Vale W, Rivier J, Thorner MO (1985) Pulsatile growth hormone secretion in normal man during a continuous 24-hour infusion of human growth hormone releasing factor (1–40). J Clin Invest 75: 1584–1590
178. Van Loon GR, Brown GM (1975) Secondary drug failure occurring during chronic treatment with LHRH: appearance of an antibody. J Clin Endocrinol Metab 41: 640–643
179. Veldhuis JD, Rogol AD, Johanson ML, Dufau ML (1983) Endogenous

opiates modulate the pulsatile seretion of biologically active luteinizing hormone in man. J Clin Invest 72: 2031
180. Waxman J (1984) Analogues of gonadotropin releasing hormone. Br Med J 288: 426–237
181. Webb CB, Vance ML, Thorner MO, Perisutti G, Thominet J, Rivier J, Vale W, Frohman LA (1985) Plasma growth hormone responses to constant infusions of human pancreatic growth hormone releasing factor. J Clin Invest 74: 96–103
182. Weiner RI, Elias KA, Monnet F (1985) The role of vascular changes in the etiology of prolactin secreting anterior pituitary tumors. In: Mac Leod RM, Thorner MO, Scapagnini U (eds). Prolactin. Basic and clinical correlates. Fidia Research Series, Liviana Press, Padova, pp 641–653
183. Weiner RI, Ganong WF (1978) Role of brain monoamines and histamine in regulation of anterior pituitary secretion. Physiol Rev 58: 905–976
184. Weissel M, Stummvoll HK, Kolbe H, Höfer R (1979) Basal and TRH-stimulated thyroid and pituitary hormones in various degrees of renal insufficiency. Acta Endocrinol 909: 23–32
185. Wenzel KW (1981) Pharmacological interference with *in vitro* tests of thyroid function. Metabolism 30: 717–732
186. von Werder K (1987) Clinical utilization of hypothalamic hormone. In: Brown G, Collu R, van Loon GR (eds). Clinical neuroendocrinology, Blackwell, Boston, in press
187. von Werder K (1985) Recent advances in the diagnosis and treatment of hyperprolactinemia. In: Imura H (ed) The pituitary gland. Raven Press, New York, pp 405–439
188. von Werder K, Fahlbusch R (1978) GnRH and TRH stimulated GH secretion in active and inactive acromegaly. In: Voelter W, Gupta D (eds) Hypothalamic hormones. Verlag Chemie, New York Weinheim, pp 677–683
189. von Werder K, Losa M, Müller OA, Schweiberer L, Fahlbusch R, Del Pozo E (1984) Treatment of metastasising GRF-producing tumour with a long-acting somatostatin analogue. Lancet 2: 282–283
190. von Werder K, Müller OA (1975) Medical therapy of hypothalamic diseases. Acta Neurochir (Wien) 75: 147–151
191. von Werder K, Müller OA (1985) Corticotropin- and growth hormone releasing factor (CRF and GRF) in the diagnosis of hypothalamo-pituitary diseases. Neurosurg Rev 8: 155–165
192. von Werder K, Müller OA, Hartl R, Losa M, Stalla GK (1984) Growth hormone releasing factor (hpGRF)-stimulation test in normal controls and acromegalic patients. J Endocrinol Invest 7: 185–191
193. de Wied D (1982) Neuropeptides and psychopathology. Eur J Clin Invest 12: 281–284
194. Williams I, Berelowitz M, Joffe SN, Thorner MO, Rivier J, Vale W, Frohman LA (1984) Impaired growth hormone responses to growth hormone releasing factor in obesity. N Engl J Med 311: 1403–1407
195. Zingg HH, Patel YC (1982) Biosynthesis of immunoreactive somatostatin by hypothalamic neurons in culture. J Clin Invest 70: 1101–1109

B. Technical Standards

Sphenoidal Ridge Meningioma

D. FOHANNO and A. BITAR

Clinique neurochirurgicale (Prof. B. Pertuiset), CHU Pitié-Salpêtrière, Paris (France)

With 23 Figures

Contents

Introduction	137
First Symptoms	139
Clinical Examination	141
Investigations	142
Plain X-Rays	142
CT-Scan	143
EEG	146
Gamma-Scan	146
MRI	147
Angiography	148
General Surgical Considerations	151
The Decision for Operation in Sphenoidal Ridge Meningiomas	155
Preoperative Care	156
Anaesthesia	157
Positioning the Patient on the Operation Table	158
Surgical Management of Sphenoidal Ridge Meningiomas	158
Management of Deep-seated (Medial) Sphenoidal Meningiomas	159
Removal of Lateral Sphenoidal Wing Meningiomas	168
Management of "en plaque" and Invasive Meningiomas	170
Postoperative Care	171
Results	171
Acknowledgments	173
References	173

Introduction

Sphenoidal ridge meningiomas originate from this sharp limit separating the subfrontal region from the temporal fossa. As Cushing and Eisenhardt[5] put it in their Monography published in 1938: "Sharply demarcating frontal from middle basilar fossa, a bony ridge curves outward on a horizontal plane from the anterior clinoidal process toward the lateral aspect of the cranial chamber where it flares out to become lost in the pterional region of the cranial vault. This landmark, commonly referred to

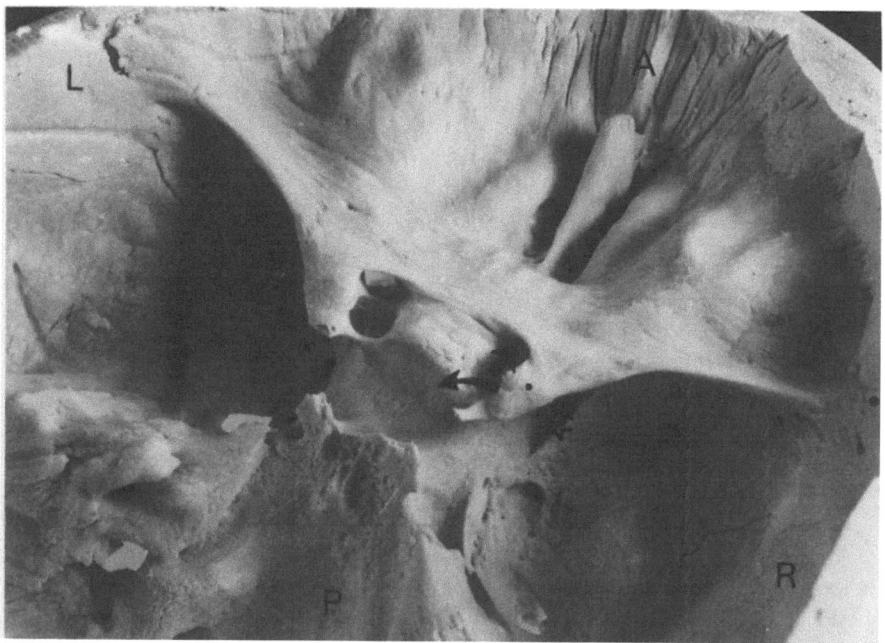

Fig. 1. ¾ oblique posterior view of the sphenoidal ridges. Large black arrow: optic nerve in the optic foramen. Curved black arrow: internal carotid artery. White arrow: supra-orbital fissure and cavernous sinus. Small arrow: sphenoidal ridge. Small dot: anterior clinoidal process. Large dot: pterional region *A*: anterior, *P*: posterior, *R*: right, *L*: left

by neurosurgeons as the "sphenoidal ridge", may conveniently be divided into three more or less equal portions: 1. deep, inner or clinoidal: 2. middle or alar: and 3. outer or pterional. The inner two thirds roughly correspond to the posterior margin of the lesser sphenoidal wing (*ala parva*): the outer third to the flaring margin of the greater wing (*ala magna*)". This description was based on anatomical, surgical and clinical considerations. With the development of new diagnostic procedures (CT-Scan and MRI) technical problems become preponderant, so that it seems more appropriate to divide these tumors into two groups: medial arising from the inner third, lateral from the external two thirds of the ridge. These two groupings pertain to globular tumors; another, third, group is related to more extensive tumors, namely "en plaque" and invasive meningiomas. The former group develops around the pterional region and provokes hyperostosis which can spread distally to invade all of the greater wing of the sphenoid and in some cases to extend beyond its limits towards the midline of the base of the skull in the sphenoid body: in these cases the meningioma itself is flat and limited to a thin layer. In the latter type tumoral tissue is globular but spreads through the temporal bone, to the superficial temporal region invading temporal muscle, and to the orbit (Fig. 1).

The sphenoidal ridge meningiomas account for 14% of all meningiomas from data found in publications dealing with this subject, and for 44 (22%) of 198 supratentorial meningiomas operated upon in the author's series. They occur more frequently in females than in males as observed for other locations, and the average age at presentation is around 50 years. The average age was 49.7 years (range 21–70 years) in this series in which only operated patients have been studied.

The difficulty of removal of these tumors varies greatly from one type to an other: for medial meningiomas the challenge comes from the relationships with the internal carotid artery, middle cerebral artery and optic nerve. For lateral ones, bleeding at the beginning of operation, removal of the bone insertion and separation of the middle cerebral artery are the main difficulties that have to be dealt with. As for "en plaque" or invading tumors the possibility and risk of total removal and permanent cure remain a subject for discussion.

In this paper the authors describe clinical findings, investigations, general consideration about surgery, decision for operation, technical requirements and results.

First Symptoms

The first symptom leading to the discovery of the tumor may be present for a very long time before it is recognized. There are many cases of patients who have had troubles progressing during several years—even more than 10 or 20—. Today most of these tumors are discovered earlier as patients have better access to healthcare. Worried by apparently mild troubles, they have a medical examination and the benefit of modern investigations. If the lapse of time between first symptom and diagnosis has tended to shorten, many meningiomas are still discovered when their size is large as they may grow without giving any trouble. In the series used in this paper mean duration between the onset of the first symptom and diagnosis was 24.5 months, ranging from 0.5 months to 20 years. The time involved was shorter for lateral tumors (14 months) than for medial ones (21.7 months). There were only 3 cases of small tumors.

Table 1. *Sphenoidal Ridge Meningiomas* (reason for consultation—41 cases)

	Medial (18)	Lateral (23)
Diplopia	3	1
Visual loss	5	1
Headache	7	9
Seizure	6	14
Psychiatric disorder	6	10
Other symptoms	6	11

The presenting symptoms (Table 1) are four: headaches, visual impairment, seizures and psychiatric disorders. Headaches are very common (39%). In some cases they are typical of increased intracranial pressure, beginning during the second half of the night, with nausea or vomiting. More frequently they are not so characteristic: mild pain in the frontal or temporal regions, occurring without any clear timing. Only sometimes do they predominate at the site of the meningioma. It is quite understandable that such headaches can be neglected for a long time as unimportant before the diagnosis of an intra-cranial tumor is even considered.

Visual impairment is much more frequent with medial sphenoidal ridge meningiomas. Unfortunately this ominous symptom is often neglected by patients. A slight unilateral blurring of vision is perceived as a small discomfort and patients only begin to worry when vision is significantly reduced after several years of evolution. When the tumor is extending to the lateral wall of the cavernous sinus diplopia may occur. It is clear that in many cases visual troubles are related to intracranial hypertension occurring at the end of a long evolution. For lateral tumors these may be the very first symptoms leading to the discovery of the lesion. Unilateral exophthalmos is observed in medial tumors or in cases of "en plaque" pterional meningiomas with bony thickening spreading to the inner part of the lesser sphenoidal wing. Invasion of the orbit by a globular invading meningioma can also happen and produce exophthalmos.

Epilepsy is a common manifestation in these patients: 50% in the authors' series, more frequent with lateral meningiomas (61%) than with medial (33%). Many different types of epileptic seizures can occur: grand mal; Jacksonian with convulsions occurring in the superior limb or the face; language suspension during several minutes or transient hemiparesis with complete recovery: so-called "temporal seizures" with the perception of abnormal, unpleasant smell, dreamy state, etc. ...

Such symptoms now lead very quickly to a CT-Scan and prompt diagnosis is usually achieved. In this localization, however, epileptic seizures seldom occur with small tumors.

Psychiatric disorders are also very frequent (40%). Major troubles with mental deterioration antisocial behaviour are rather uncommon and observed with huge tumors or meningiomas arising from the sphenoidal ridge and extending over the orbital roof under the frontal lobe. Minor troubles are common: decrease of power to concentrate, impairment of memory, lack of interest in customary activities, anxiety irritability are frequently found. In many cases they are neglected by the patient himself and may only be discovered by the examination of his professional records or through a thorough inquiry of his family, or people living by his side.

Other presenting features are: paresthesia in the face, impotence, urinary incontinence in medial tumors; aphasia, vertigo, sexual impotence and urinary incontinence, with hemiparesis in lateral placed lesions.

Clinical Examination

In a schematic presentation the details of clinical examination in sphenoidal meningiomas are directly related to their localization. The medial sphenoidal ridge meningiomas involve the angle between the clinoid process and the cavernous sinus where the optic nerve goes into the optic foramen and the oculomotor and fifth nerves go through the supraorbital fissure. Bone thickening with compression of orbital venous drainage also contributes to produce ophthalmological symptoms—ipsilateral visual loss, homonymous hemianopsia, ophthalmoplegia, corneal hypoesthesia and exophthalmos. By contrast, lateral meningiomas give rise to nonspecific symptoms such as mild headaches as they are related to nonfunctional areas especially in the non dominant hemisphere. Seizures can be elicited by fronto temporal disturbance due to compression or irritation. Invasive tumors produce visual involvement and exophthalmos. In fact clinical presentation is much more complex as the slow and silent growth of many of these tumors—especially lateral—leads to signs of raised intracranial pressure: headaches, visual impairment psychological changes which are nonspecific.

The results of clinical examination in the patients we present here are summarized in Table 2. (The results for patient with "en plaque" or invading meningiomas have not been included as there are only three of them, too small a number to have any statistical value.)

On neurological examination increased deep reflexes, hemiparesia speech impairment are the most common symptoms (40%) patients with a clear predominance in favor of lateral tumors. Unsteady gait is also observed. Facial hypoesthesia have been seen with medial tumors (11%). It must be stressed that 34% of patients had a normal neurological examination.

The proportion of patients with normal ophthalmological examination

Table 2. *Sphenoidal Ridge Meningioma* (result of clinical examination—41 cases)

	Medial (18)	Lateral (23)
Neurological symptoms		
hemiparesia—speech impairment	1	19
increased reflexes	4	2
unsteady gate	1	3
facial hypoesthesia	2	0
normal	4	10
Ophthalmological symptoms		
visual loss	6	3
oculomotor palsy	3	1
exophthalmos	1	0
fundi alteration	9	8
visual fields changes	4	3
normal	7	14
Foster Kennedy syndrome	3	

is also great: 51%, with no significant difference between the two groups. The main troubles are visual acuity loss (22%) predominant in medial tumors (33% of the localization), oculomotor palsy (10%), exophthalmos (1 case), visual field changes (17%) and abnormal fundoscopy: either haziness of the disc margin, papilloedema or optic nerve atrophy (42%, equally distributed, with 3 cases of Foster Kennedy syndrome related to medial tumors).

Investigations

Many investigations can be performed. They are described here but it must be clear for both medical and economic motives that all of them are not used in the same patient, as will be discussed later. It is necessary to bear in mind that the two aims are 1. to get an unquestionable diagnosis of the presence of a tumor, its location and size and 2. to obtain before operation an idea as accurate as possible about its insertion on basal dura, relation to main vessels and vascularization.

1. Plain X-Rays and tomography are the first necessary investigations. The former being routinely made, the latter only when a doubt about a small change occurs on plain X-Rays. The main change is thickening and increased density of the lesser wing seen on antero-posterior views (Fig. 2). Thickening of the squama of the temporal bone may be seen on lateral view as a condensation of the pterion (Fig. 3) (the so-called "smoking pterion").

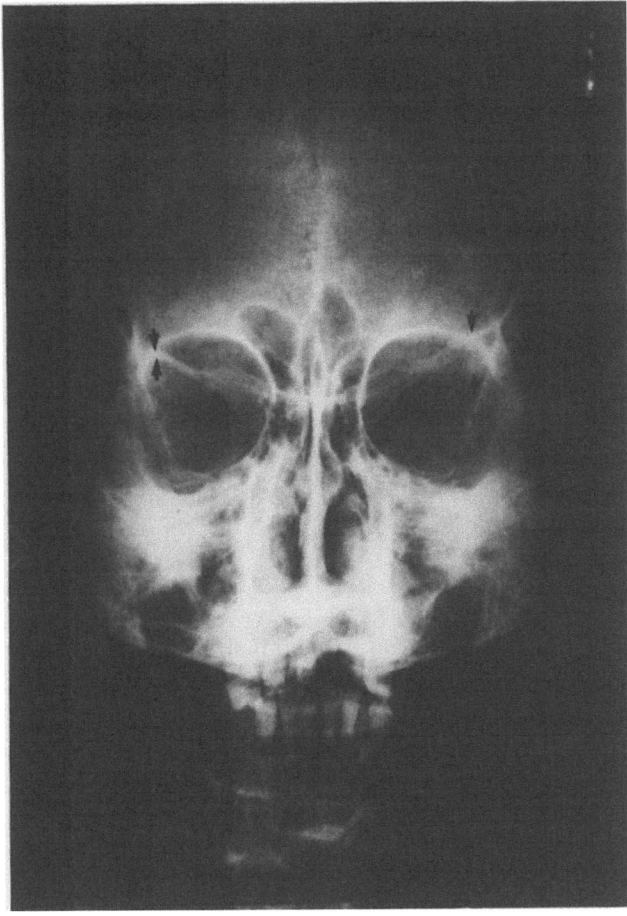

Fig. 2. Plain X-Ray study of lateral left sphenoidal ridge meningioma. Thickening of the pterion and lesser sphenoidal wing. Frontal view

The authors have seen many patients in whom a plain X-Ray had been made many years before for uncharacteristic symptoms and which had been misleading owing to inadequate technique—the lesser wing must be well defined in the orbit—or poor interpretation. Tomography yields indisputable evidence when bony changes are present (Fig. 4).

2. CT-Scan is now the indispensable and determinant investigation. It is mandatory to make pictures with and without contrast injection for, in some examinations without injection of contrast, the meningioma may not be visible. With contrast enhancement the tumor appears clearly and when using appropriate window settings the changes of bone density are obvious (Figs. 5, 6, 7). Standard axial view scanning of the whole brain, including the posterior fossa is necessary as some patients have several meningiomas.

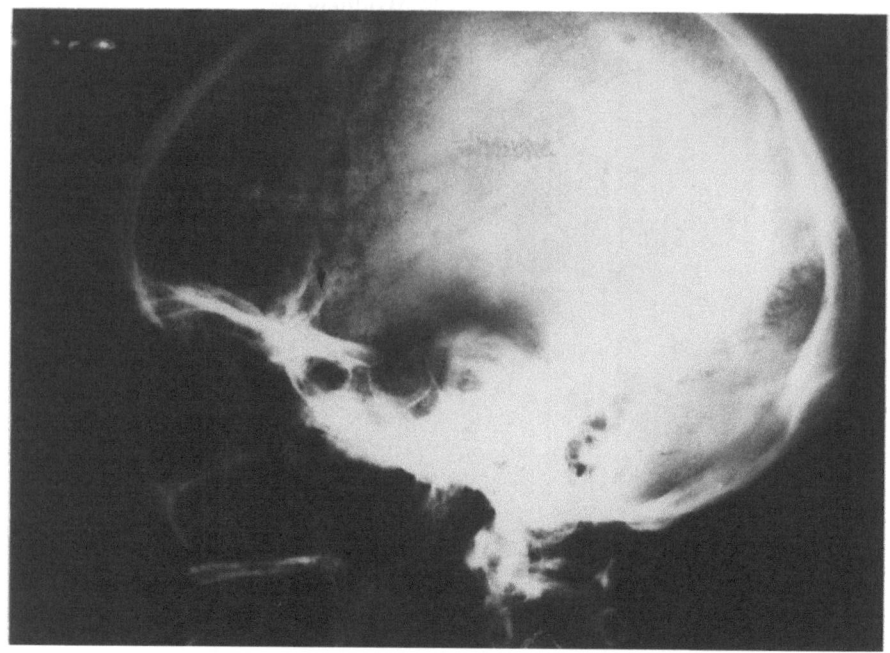

Fig. 3. Plain X-Rays (lateral view) of pterional meningioma showing increased density of the bone at pterion

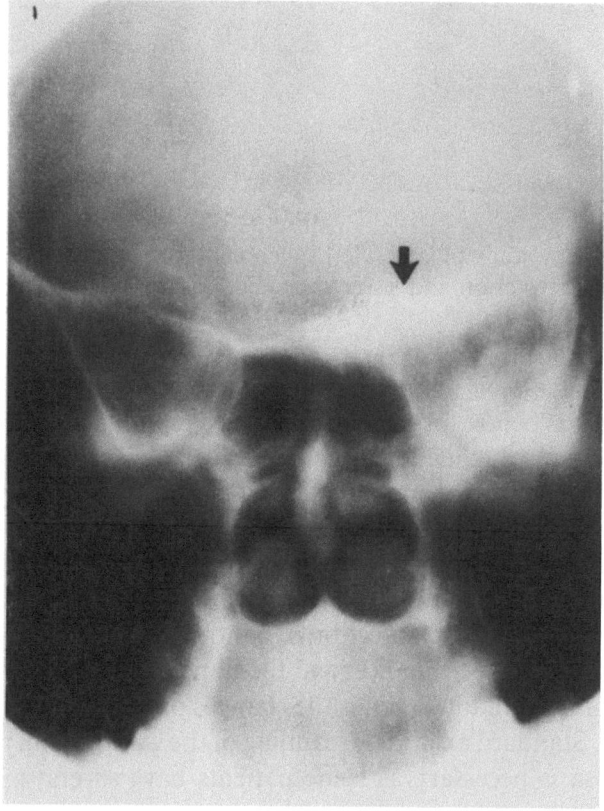

Fig. 4. Frontal tomography of an "en plaque" meningioma of the sphenoidal ridge. Note the thickening of the bone of the roof and of the lateral wall of the orbit

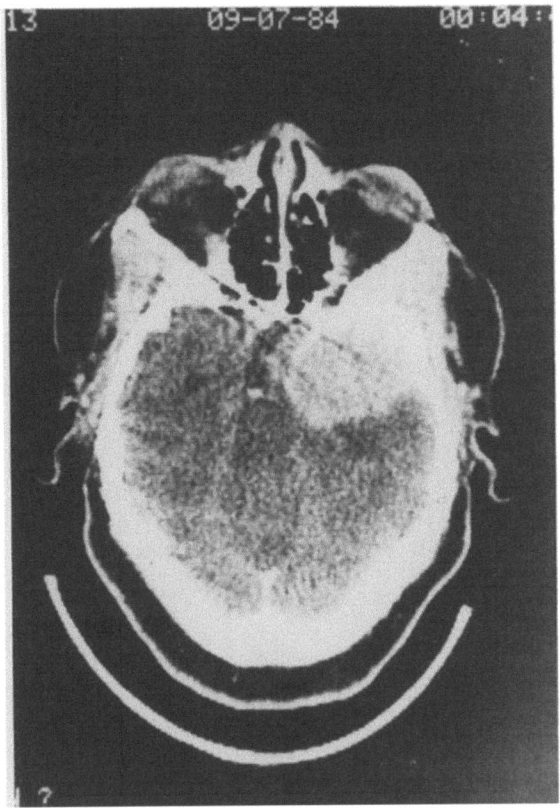

Fig. 5. Axial CT-Scan view of a medial sphenoidal meningioma developing in the temporal fossa

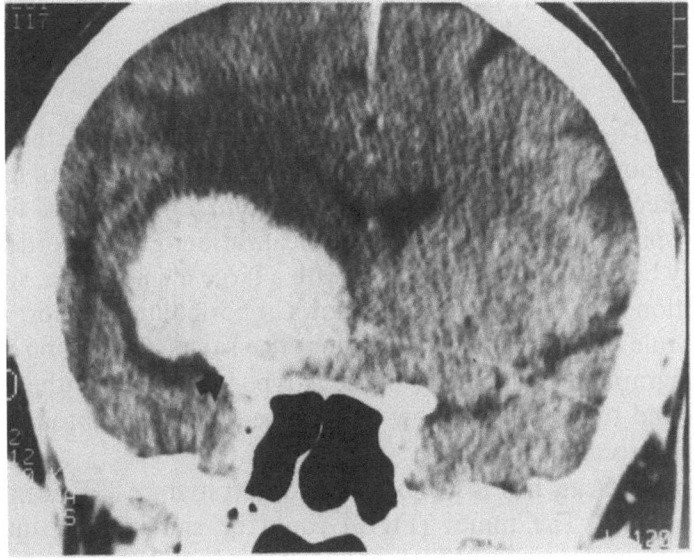

Fig. 6. Medial (clinoidal) meningioma of the sphenoidal ridge. CT-Scan. Frontal view. Enhancement after contrast injection. Black arrow: thickened anterior clinoid process

With injection and appropriate sections, carotid artery and middle cerebral artery can often be displayed and their relation with the tumor shown. The clearest representation of these sphenoidal ridge meningiomas, particularly the medial ones, is obtained with coronal sections. With small tumors serial thin slices are necessary and must be repeated some weeks or months later if they are not conclusive.

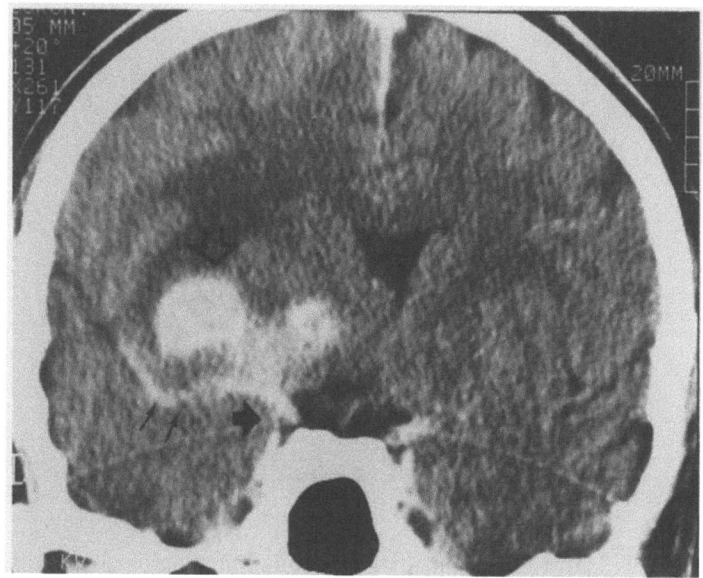

Fig. 7. Same patient as in Fig. 6. Clinoidal meningioma. Relation with internal carotid (large black arrow) and middle cerebral (small black arrows) arteries. Open arrow: meningioma

3. Electro-encephalography and Echo-encephalography: there are two kinds of abnormal change in the EEG: slow waves localized to the fronto-temporal region on the side of the meningioma and epileptic features of spikes or spikes and waves in the same localization. EEG records may also frequently be normal, however, even with a large meningioma, so that this investigation is useful only when it shows abnormalities. A normal record does not rule out an intracranial tumor. The same observation is true for echotomography, for only echo of the A-type is available without opening the skull and this can only show a "midline echo", so that many tumors remain undetected.

4. Gamma-Scan has lost most of its interest since CT-Scans became easily available. The tumor (Fig. 8) could be shown after intravenous injection of radioisotope (*Technetium Pertechnetate*) with its size and

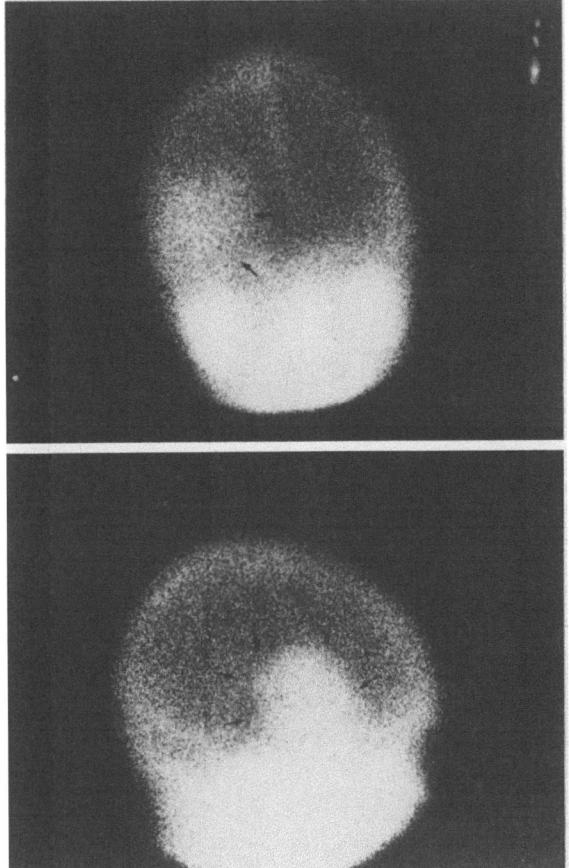

Fig. 8. Pterional meningioma. Gamma-scan after injection of *Technetium pertechnatate*. Anterior and lateral view

localization. But visualization of surrounding structures was impossible and this investigation is now out of date. Its major advantages were the possibility of study of the type of persistence of isotope which was different with malignant tumors (late and long lasting) and meningiomas (early uptake disappearing in few hours), and its safety. The main disadvantage is that it is unable to differentiate small tumors (under 2 cms in diameter) from the normal uptake of isotope at the base of the skull.

5. Magnetic resonance imaging (MRI) probably has fine prospects before it because the abnormal signals (Figs. 9, 10) show the meningioma, the oedema which surrounds it and the main vessels. It cannot be surpassed in the field of non-invasive investigations when it becomes necessary to see the sagittal arrangement of a tumor and its relations with midline or near-midline anatomic structures.

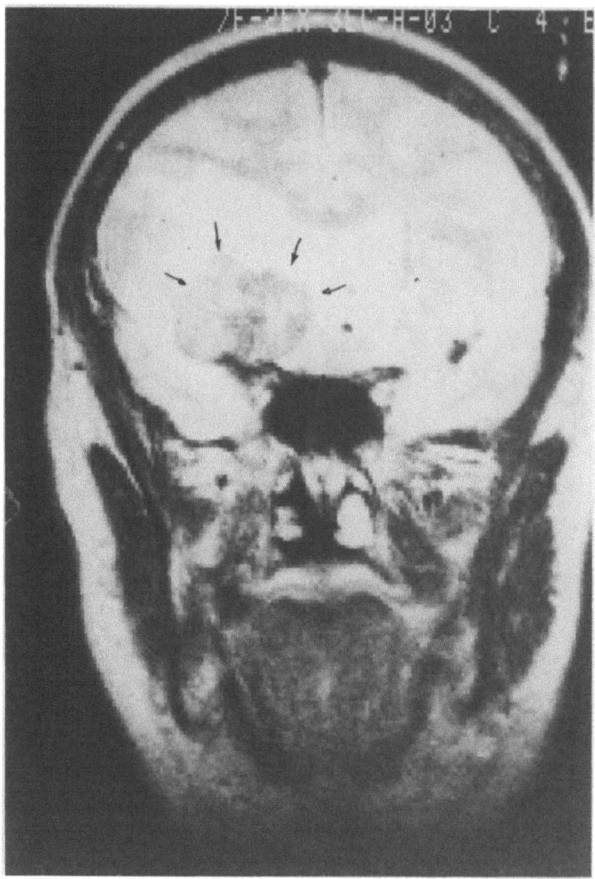

Fig. 9. Clinoidal meningioma. Magnetic resonance imaging. Same patient as in Figs. 6 and 7. Frontal view

The other investigations are not harmless and may be painful and hazardous for the patient. Air ventriculography, used for many years by neurosurgeons, is too dangerous and painful to be used any more. Air encephalography which could be useful to delineate small tumors near the midline some years ago has now been totally replaced by MRI or advanced CT.

6. Angiography remains necessary. In most cases global angiography with separate injection of the internal and external carotid arteries will yield the necessary data (Figs. 11, 12): size and vascularity of the tumor, arterial supply, relations with the major blood vessels. This most important point is not easy to specify. In many cases internal carotid and middle cerebral arteries are only displaced but their diameter remains unchanged and there is no stenosis; in these cases the vessel is only displaced by the tumor without

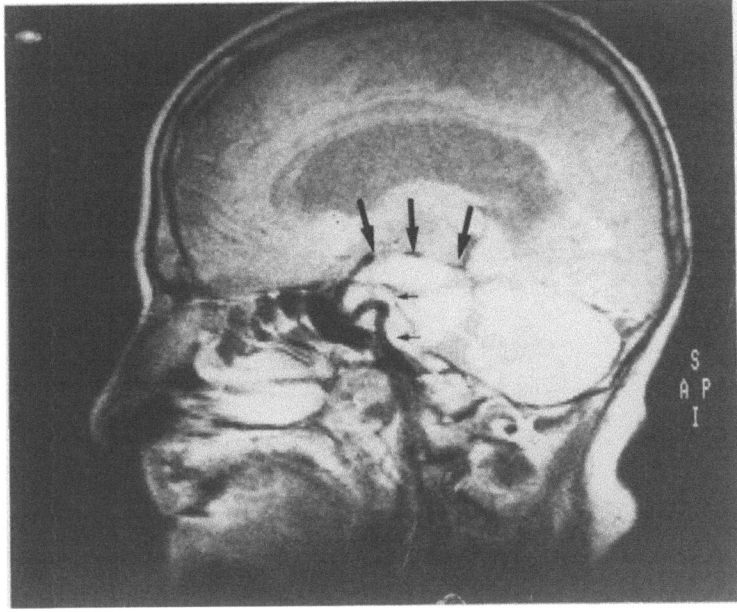

Fig. 10. Medial expansion of a medial sphenoidal ridge meningioma (sagittal section). (MRI) Large arrows: meningioma, small arrows: internal carotid artery with low signal due to velocity of blood flow

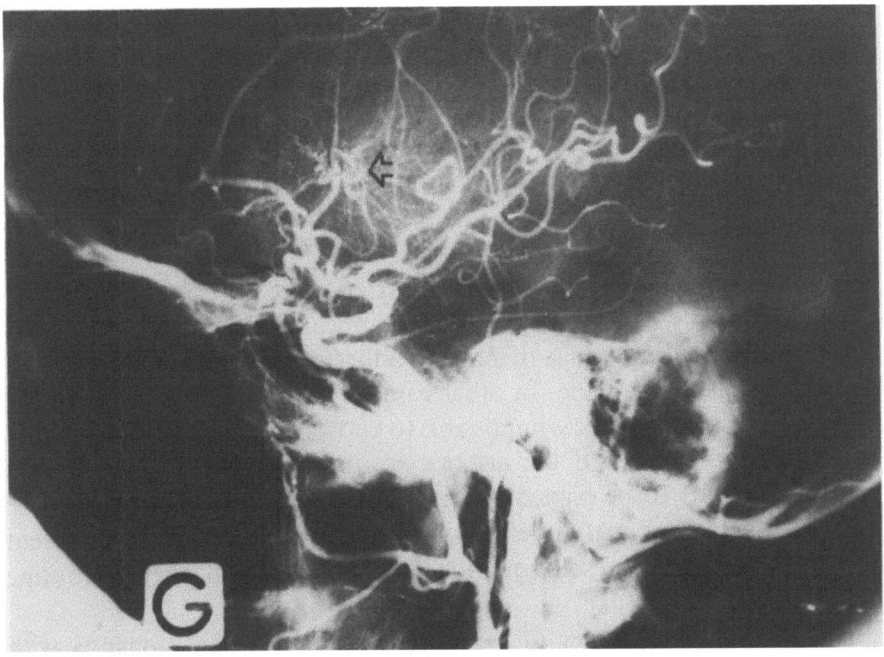

Fig. 11. Angiography of a lateral (pterional) meningioma. Lateral view. Black arrow: dilated medial meningeal artery penetrating (open arrow) the tumor

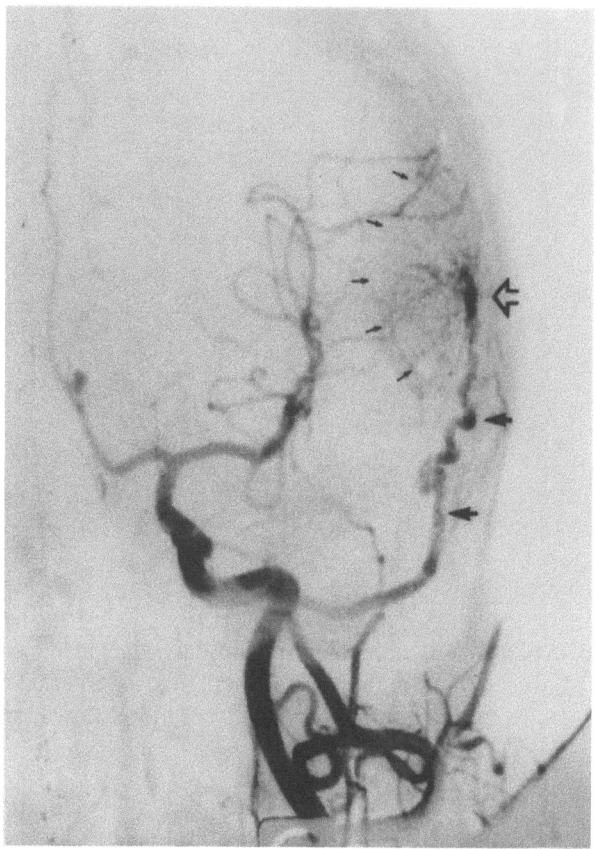

Fig. 12. Angiography of a pterion (lateral) meningioma. Small arrows: internal edge of the tumor. Black arrows: feeding vessels coming from the external carotid artery. Open arrow: penetration of feeding vessels into the meningioma

direct contact. In most of these cases the arteries are hidden and protected by the arachnoid which during operation appears thickened and loses its normal transparency, so that there is no adhesion to the meningioma and the dissection is easily made. In other cases these arteries are streched the normal curve disappears. It is impossible to know before operation, whether dissection is possible or if the arteries are in the tumor with their wall invaded. The part to be played by digital angiography cannot clearly be known as the lack of spatial definition of this investigation in the present state of the technique does not provide all the information given by routine angiography.

The general practitioner or neurologist whose patient consults with symptoms consistent with intracranial tumor and specifically sphenoidal meningioma would obtain plain X-Rays of the skull and arrange referral for

CT-Scan. Early investigation of this kind is much better even if most of these examinations are negative. The descovery of a small sphenoidal ridge meningioma is better than waiting and referring the patient to the neurosurgeon with a tumor which has grown too big to be totally removed without great risk of functional impairment or vital hazard. Further investigations must be decided only in the perspective of an operation, to delineate the anatomical details precisely so that the surgeon can intelligently plan the operation.

General Surgical Considerations

Sphenoidal ridge meningioma still remains a challenge to the neurosurgeon. The safe resection of this tumor to obtain complete and permanent recovery would imply detection 1. of small tumors. 2. Suppression or at least significant decrease of tumor blood supply during the first steps of operation before beginning tumor removal 3. total removal of the meningioma including the dural insertion and involved bone 4. avoidance of major damage to the brain and its supplying vessels.

All these requirements are not possible for most patients.

External carotid artery ligation was proposed some seventy years ago to diminish the often tremendous bleeding of pterional meningiomas. More sophisticated and more specific procedures using embolization have been proposed recently[18]. They do not afford a completely satisfying solution as the bleeding can be controlled at the beginning of operation with pterional meningiomas as described later while branches of the internal carotid artery supplying the medial tumors cannot be controlled in this way though it might be very useful: direct access to the feeding vessels of the tumor is not available before the end of the operation when the last part and the attachment of the tumor is removed.

Complete ablation is impossible in some cases. Pterional "en plaque" meningiomas may extend very far to the midline and total removal is impossible when bony invasion involves the supraorbital fissure, cavernous sinus, foramen spinosum and, under the base of the skull, the pterygoid process. This may also happen with invasive meningiomas (Figs. 13, 14).

Total removal is limited in medial meningiomas by their relation to the internal carotid and middle cerebral arteries (Figs. 15, 16, 17). When they are included in the tumor total removal with a good functional result is impossible. When adhesions between the tumor and these arteries are found, total extirpation is theoretically possible, but things look different when considering the results. A technically perfect operation is not always the best for the patient. A number of cases have been reported in the literature—and we have had or heard of several examples—in which total removal of tumor has been obtained by cautious dissection of the Sylvian

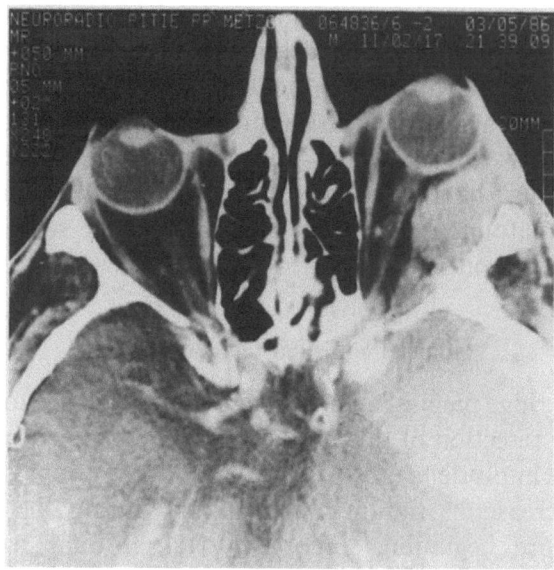

Fig. 13 a

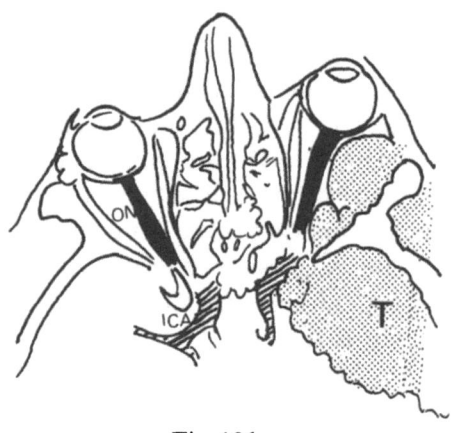

Fig. 13 b

Figs. 13 a, b. Axial CT-Scan view of a large globular, invasive sphenoidal ridge meningioma showing the intra-cranial and intra-orbital extensions. Note the relations with the internal carotid artery and the optic nerve. *T*: tumor, *ON*: optic nerve, *ICA*: internal carotid artery

vessels or internal carotid artery under the micoscope with no surgical lesions of these vessels and angiograms performed after operation were normal or showed mild spasm only. Weeks or even months later, repeat angiography, sought because of occurrence of a hemiparesis, displayed complete thrombosis. It is probably better to leave behind a small piece of tumor in contact with the vessels than to take such a risk. The future does

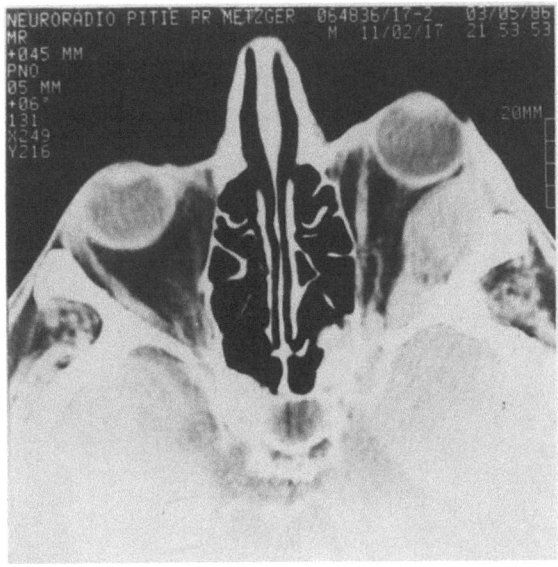

Fig. 14 a

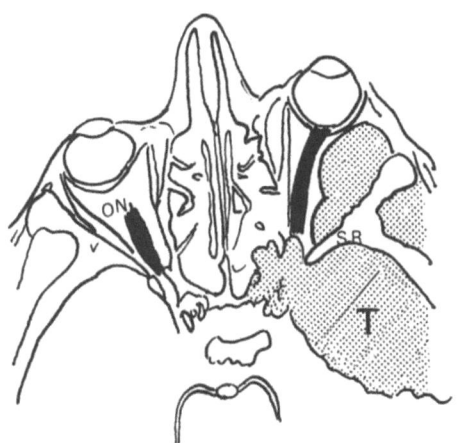

Fig. 14 b

Figs. 14 a, b. Same patient as in Fig. 13. Relation between the tumor and optic foramen. *SR*: sphenoidal ridge

not seem to be impaired when these remnants have been separated from their insertion. When relations with the insertion remain the chance of later regrowth is always possible accounting for the high number of recurrences of meningiomas with a medial insertion.

Three other aspects are of the utmost interest for the surgeon as they may make for a relatively easy or extremely difficult operation: namely, the

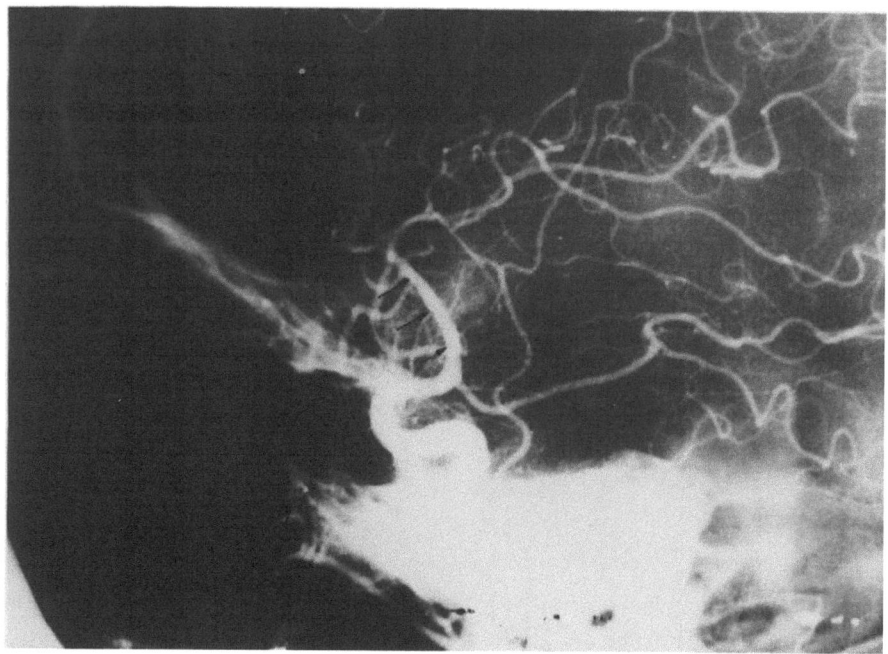

Fig. 15. Medial sphenoidal ridge meningioma. Lateral view of angiography. The small arrows show the stretched internal carotid and Sylvian arteries. Note the irregularities of the wall of the carotid artery which was embedded in the tumor at operation

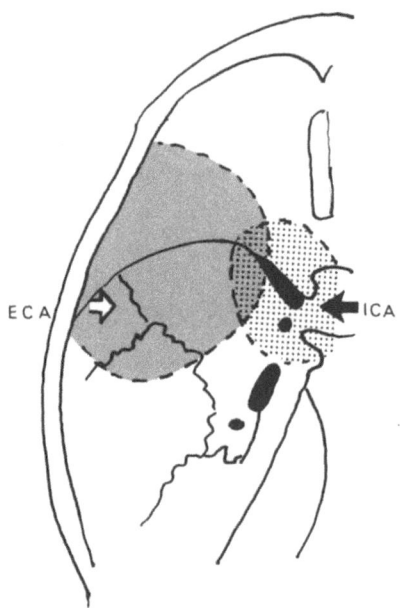

Fig. 16. Schematic drawing of vascularization of sphenoidal ridge meningiomas. *ECA*: external carotid artery. *ICA*: internal carotid artery

importance of vascularization of the tumor and the possibilities of early control during operation, firmness or looseness and extension of its insertion. Difficulties are maximal with a vascular hard tumor with a large base of insertion whereas a tumor with a narrow insertion on the sphenoidal ridge, a loose consistency and which does not bleed much is easier and safer to remove.

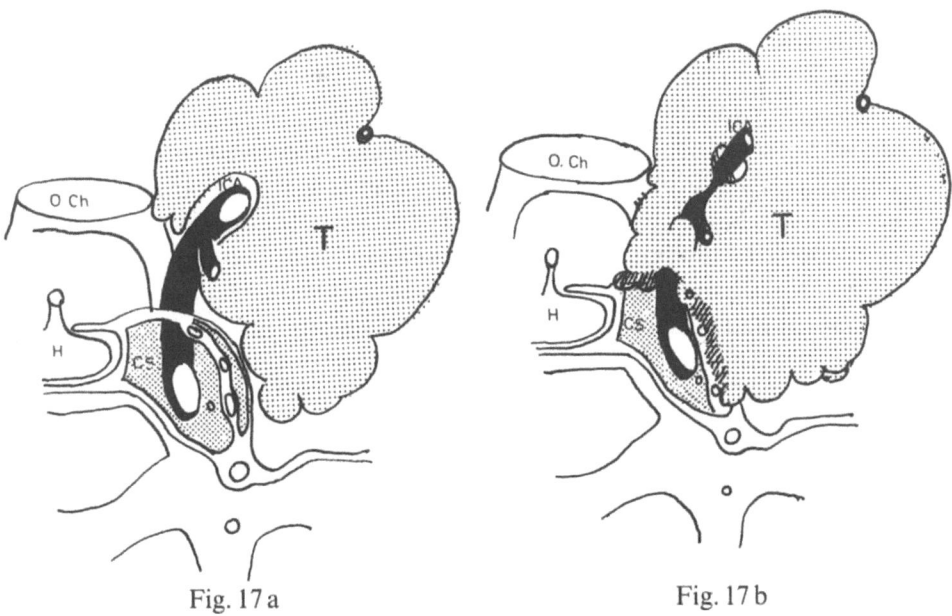

Fig. 17 a   Fig. 17 b

Figs. 17 a, b. Relations between medial sphenoidal ridge meningioma, internal carotid artery and cavernous sinus (frontal section). *T*: tumor, *O Ch*: optic chiasma, *H*: hypophyseal (pituitary) gland, *CS*: cavernous sinus, *ICA*: internal carotid artery. a) The cavernous sinus and carotid artery are surrounded by the tumor and dissection is possible. b) Cavernous sinus and carotid artery invaded by meningioma. Total removal is impossible

The Decision for Operation in Sphenoidal Ridge Meningiomas

It is easy to understand for the preceding reasons that the decision for operation must be thoroughly thought over. The surgeon must not be pusillanimous but not foolhardy. He will have to be cautious when deciding to operate and in planning the operation. He must be able to change his behaviour during the operation if the local situation is different from what he had anticipated. When making his decision for operation he must remember that these tumors are slow growing ones and the authors agree with the opinion of Clark Watts[19] that "fifteen years of gradually decreasing vision and ophthalmoparesis, punctuated by two or three operative procedures, may be much better for the patient than fifteen years of blindness and complete ophthalmoplegia without recurrence following a single radical surgical procedure".

This subject has been discussed by Pertuiset et al.[15]. Some patients must not be operated on: those with major medical diseases such as lung, heart, kidney impairment, diabetes mellitus, for they are unable to survive a shaky postoperative course, a frequent condition with huge or medial meningiomas. It is also impossible to remove medial tumors occurring in patients over 70 years of age. Lateral tumors may be operated but attempts to accomplish total removal would be inadvisable.

For these patients a better management seems to be to perform partial removal of the lesion. Through a small craniectomy localized at the fronto-temporal region it is possible to avoid the hazards of dural tearing, cortical laceration and lesions of the Sylvian veins that may occur in elderly patients when separating the bone flap from the adherent dura; coagulation of the tumoral insertion is carried out, dura opened and tumor gutted out with curettes—or better, ultrasonic aspiration—leaving the capsule of the tumor in place.

Preoperative Care

It is advisable to control brain oedema and the risk of epileptic seizures before operation. Limitation of bleeding by external carotid ligation or embolization needs to be considered. The authors routinely use corticoids to decrease peritumoral oedema. This is mandatory for patients with clinical symptoms of raised intracranial pressure or when a large area of hypodensity surrounding the tumor is present on CT-Scans. Four injections per day of 4 mgs of dexamethasone is a standard dose. For those patients with peptic ulcer drugs protecting the gastric mucosa must be added (Eg Cinetidine 200 mgm. Three Times day). Unless the peptic ulcer is clearly active, the advantages of steroids outweight their potential hazards.

Systematic use of antiepileptic drugs such as phenobarbital is practically without danger when looking at the risk of epilepsy and we think it better to use such a treatment in all patients during the preoperative period.

Many neuroradiologists or neurosurgeons have tried to reduce bleeding at operation by controlling the blood supply to the tumor. The benefit of suppressing most of the arterial pedicles going to the tumor has been emphasized by R. Djindjian promoting a sophisticated method of supraselective embolization, and Teasdale[18], who advocates a simpler technique of subselective embolization. This author comparing angiographic studies, dynamic radioisotope brain scans, pathological examinations and operative findings considered that embolization significantly reduced bleeding in 13 of 26 patients who had had preoperative embolization. He stressed that most of these patients had purely or predominantly feeding vessels coming from the external carotid artery. We however have been somewhat disappointed with this method. When operating on lateral tumors an epidural approach with coagulation of meningeal vessels going to the tumor allows a quick control of bleeding. The bleeding is important but does not last very long when appropriately and quickly handled. After embolization bleeding from the main feeders is reduced but diffuse bleeding from small arteries is troublesome and the time during which this profuse bleeding occurs is longer. Real help would come from control of abnormal feeders from the internal carotid artery in medial meningiomas particularly when the tumor, as is so often the case in this localization is vascular and of hard consistency. Embolization in the internal carotid territory however cannot yet be performed without major hazard for the patient.

Anaesthesia

Many types of anaesthesia may be used during these operations. We have felt it wiser to describe the technique used by the anaesthesiologists who have taken the author's patients in charge, than to try to make a critical review which is beyond our limit. Narco-neuroleptanalgesia has been the common method used: an association of an analgesic drug such as Phenoperidine or Fentanyl, another providing narcosis like Thiopental and a third affording neuro-vegetative protection (Droleptan is commonly used). Controlled ventilation is mandadory with these drugs as they induce respiratory depression. Control of intra-arterial pressure, blood gases is essential in these procedures, when major bleeding may occur.

Several aspects deserve particular consideration:

1. Raised intracranial pressure must be controlled before opening the dura as the brain may swell suddenly. Such herniation may be compressed at the margin of the dural incision, with vascular strangulation jeopardizing chances of a good result at the very beginning of the operation. Several ways of decreasing intracranial pressure are available. The quality of anaesthesia is of major importance. Correct use of controlled ventilation with the addition of a negative phase has a determining value: lowering of PCO^2 lowers cerebral blood flow and induces venous contraction reducing intracranial mass. Positioning of the patient on the operation table is most important and care must be taken to lower intracranial venous pressure by placing the head above the level of the heart to favour venous drainage and to avoid any compression of the jugular veins.

2. Intravenous Mannitol has been used for many years to lower intracranial pressure. It is a very effective drug but it is not easy to handle as the profound fall of intracranial pressure obtained some minutes after injection may be followed by a rebound of intracranial hypertension some hours later. Increase of bleeding due to fragility of small cortical blood vessels under Mannitol has led us to limit its use to emergency cases, as brain

swelling may be avoided with modern technics of anaesthesia: namely the use of intra-venous Thiopental.

3. The control of blood pressure is generally obtained by the anesthesia described above. A stable mean blood pressure of 70 mms of mercury is recommended. Should profuse bleeding occur and be uncontrolable for a long time, as may happen with vascular medial meningiomas the blood supply of which cannot be controlled before the last stage of operation, Sodium nitroprusside may be very useful to obtain deep hypotension under 50 mms of Hg quickly during this difficult stage of the operation. Anaesthesic halogenated gases such as Ethrane or Forane may be helpful, but their use is limited by the increase of intra-cranial pressure they are liable to induce.

A great quantitiy of blood must be available at the beginning of operation as bleeding may be profuse during the skin incision and the making of the bone flap. This is a particular problem with pterional meningiomas.

Positioning the Patient on the Operation Table

The patient is operated on in the supine horizontal position with the head raised about 20° to help decrease venous pressure. The head is turned about 30° from the midline looking towards the opposite side. The ipsilateral shoulder may be slightly elevated to help turn the head without impairing the venous drainage. The neck is then slightly flexed, extended or left in a neutral position depending on the main extension of the tumor either to the anterior or temporal regions. Extension allows better visualization and access to the roof of the orbit and flexion to the temporal fossa. Increased rotation gives a better view of the lateral part of the sphenoidal ridge while a lesser degree of rotation gives more comfort when dealing with a medial tumor. It is very useful to be able to change the head positioning slightly during different stages of the operation. Using an operated table which can be tilted is a good solution. When it is not available positioning of the patient's head must be planned to allow the best exposure during the most difficult stages of the operation (Fig. 18 a).

At this time, before beginning to operate, it is necessary to check the position and to adjust the operating microscope particularly when it is implanted in the roof of the operating room to avoid later major change in the positioning of the patient during operation.

Surgical Management of Sphenoidal Ridge Meningiomas

The removal of such a tumor will be described in detail for a deep-seated medial meningioma. Most of the dangers and difficulties are encountered with this localization. More specific technical problems proper to pterional and "en plaque" extensive or invading meningiomas will be considered in two shorter sections of this chapter.

Management of Deep-seated (Medial) Sphenoidal Meningiomas

1. *Scalp incision* (Fig. 18 b): A curved fronto-temporal incision with anterior concavity starting just in front of the ear and above the zygomatic process remaining just the behind the hair line allows a good exposure to fashion the bone flap. It must not be extended further on at its anterior part to avoid an unaesthetic scar in the forehead. It is largely sufficient to give access to the frontal base, sphenoidal ridge, midline and anterior part of the middle fossa. Should a large bulk of the tumor extend in this fossa, a "Y-shaped" incision yields better access to its posterior part. Preventive coagulation, division and ligation of the dilated superficial temporal artery is required and care must be taken not to cut this artery with the knife during incision. Skin and galea may then be reflected. The authors favor infiltration of subcutaneous layers with saline as this allows easier dissection of the epicranium and superficial temporal aponeurosis from the skin. Adrenalin is added to reduce skin bleeding before it can be controlled by the usual means.

2. *Performing the craniotomy* (Figs. 18c, d): A fronto-temporal bone flap performed around the pterion gives access to all of the region, allowing a choice between lateral sub frontal, anterior temporal and/or an approach along the sphenoidal ridge. We favor the last one. We recommend a large bone flap to facilitate operative handling taking into account brain bulging due to the size of the tumor and brain oedema. It is made with five burr holes. The "keyhole" is placed at the supero-anterior part of the temporal fossa just under the anterior insertion of the temporal muscle, behind the lateral orbital pillar. A Gigli saw or any appropriate means is used to cut the bone between the burr holes. Care must be taken not to tear the dura with the guide when separating it from the bone. The danger is real with patients over sixty as adhesion between the bone and dura may be very firm. When this situation occurs a craniectomy should be performed, by removing the bone remaining between burrholes with a rongeur. Bleeding is controlled by application of bone wax. The craniotomy is finished by removing some of the bone left between the two inferior holes with a rongeur and breaking the flap at this place. The bone flap is then turned outside, most of the temporal muscle is separated from the external aspect of the bone to reduce bleeding. Bone wax is applied all around the section of the bone; dural vessels are coagulated. If necessary, haemostasis of the superficial layers, bone and dura is achieved by coagulation of muscle, and peripheral dura, application of Oxycel in the epidural space and eventually placing stay sutures between dura and epicranium. Haemostasis must imperatively be perfect at this stage of operation. The dura cannot be opened before bleeding has been

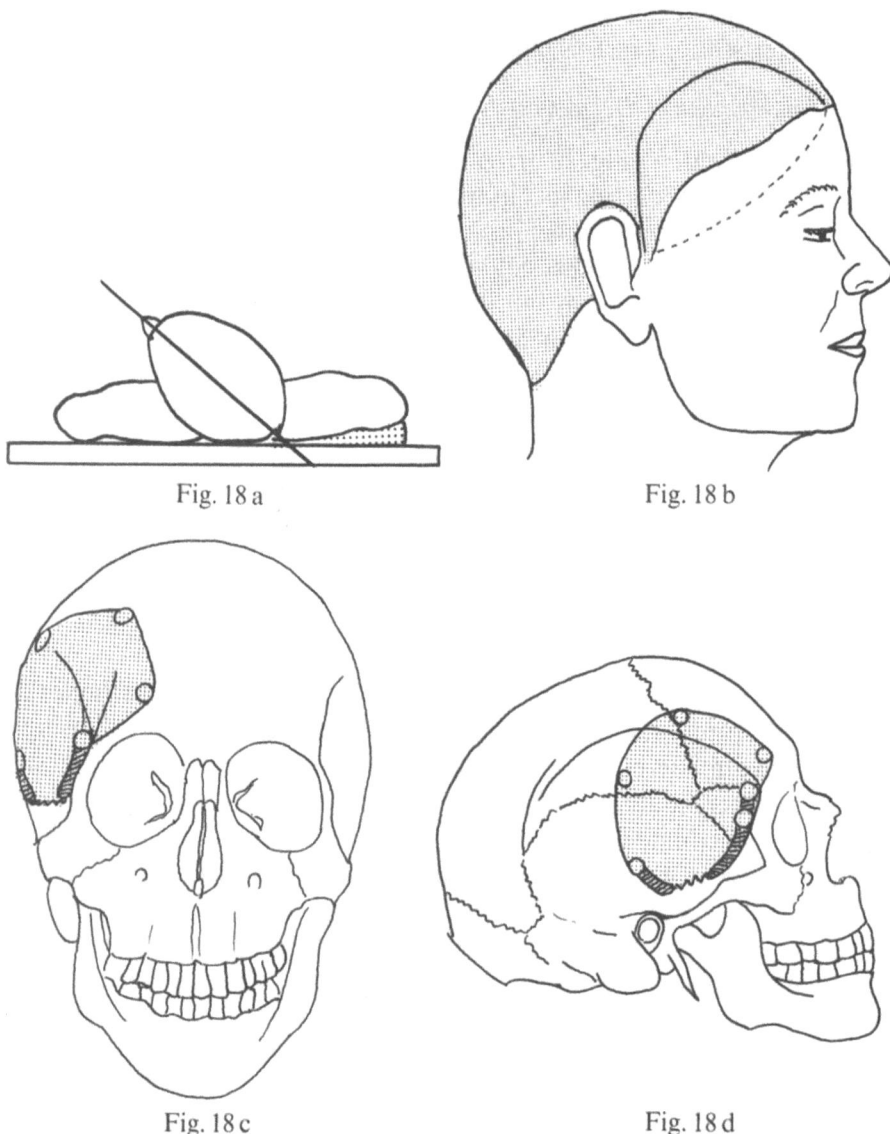

Figs. 18 a, b, c, d. a) Head positioning. b) Skin incision. The dotted line corresponds to the antero-inferior separation of the scalp from the skull. c) Bone flap, frontal view. d) Bone flap, lateral view

totally controlled. To decrease bleeding further the dura must be retracted as far as possible inwards to the edge of the tumor attachment as proposed by Cook[4].

3. *Dural opening* (Fig. 19 a). The dura is then opened around the pterion. When the meningioma is small, the brain slack and with the help of optic

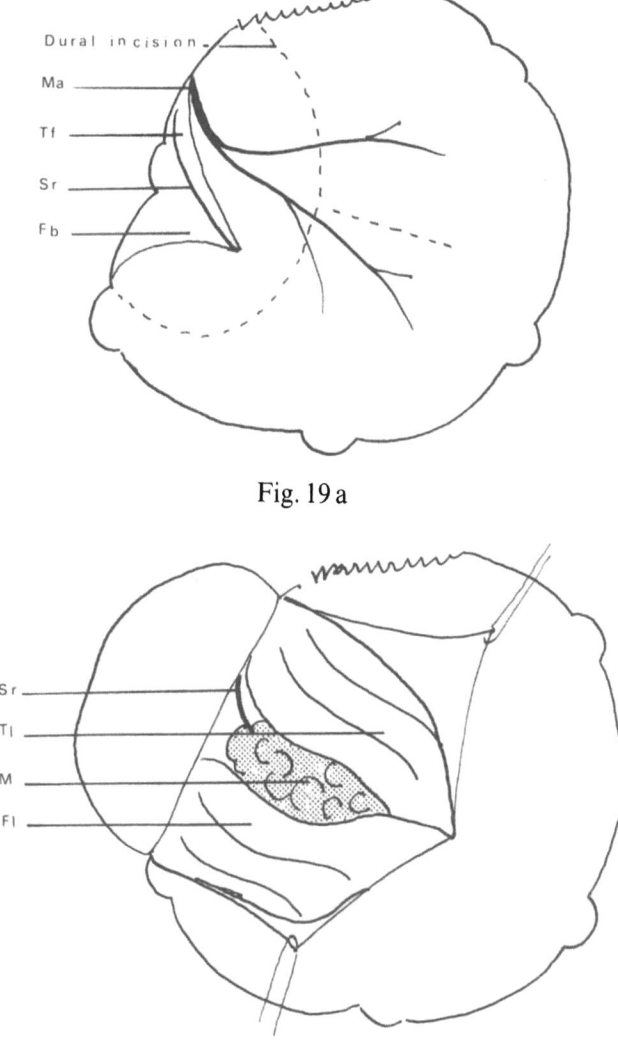

Fig. 19a

Fig. 19b

Fig. 19. Removal of a right medial sphenoidal meningioma. The drawings are placed in the position of the patient's head at operation: the upper part of the drawing corresponds to the base of the skull. *Ma*: middle meningeal artery, *Tf*: dura of temporal fossa, *Sr*: sphenoidal ridge, *Fb*: frontal base. a) Dural incision. b) Opening the dura. The tumor capsule may be seen between the frontal lobe (inferior part of the drawing) and the temporal lobe (superior part). c) Incision of the lateral aspect of the capsule and aspiration with ultrasonic aspirator. *Cot*: Cottonoids to protect the cortex. *Da*: Dural attachment of the tumor partially removed and coagulated at the lateral part to partially control bleeding. *MCA*: Middle cerebral artery hidden (dotted black thick lines) by the tumor. Only the branches of division can be seen at this stage of operation. d) Lateral part of the tumor removed. The retrograde dissection of the Sylvian artery is shown with the black curved arrow. e) Presentation (grey box) of the remaining internal part of the capsule. f) Detail of the preceding drawing (corresponding to the bose): Resection of the last piece of capsule. *Fl*: frontal lobe. *Tl*: temporal lobe. *On*: optic nerve. *Ca*: carotid artery. *Cp*: anterior clinoidal process

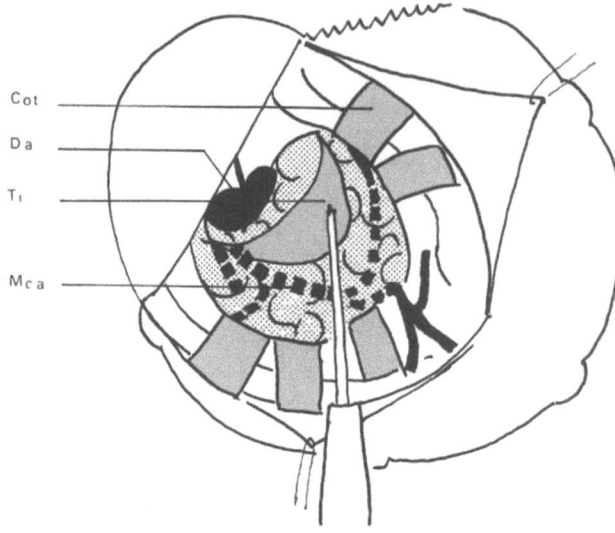

Fig. 19 c

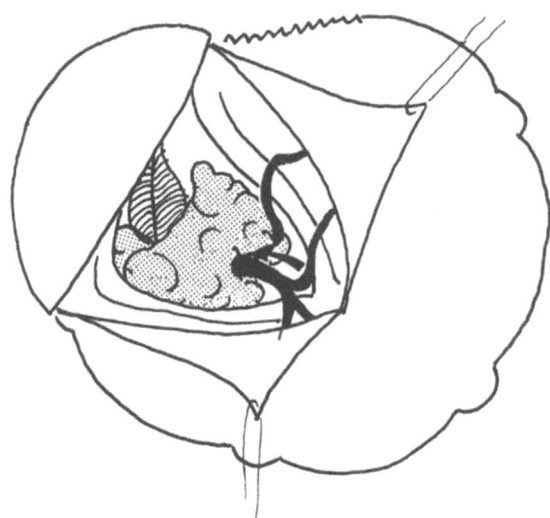

Fig. 19 d

magnification this small opening may be adequate. When the tumor is large or the brain is swelling it is recommended to branch another posterior incision to prevent brain strangulation and afford good exposure of the Sylvian fissure, inferior frontal convolution and the anterior half of the temporal lobe. Bleeding from the dural edge is easily controlled by coagulation with a forceps used with low intensity to prevent retraction. We have not been obliged to use clips on the dura to control this bleeding for

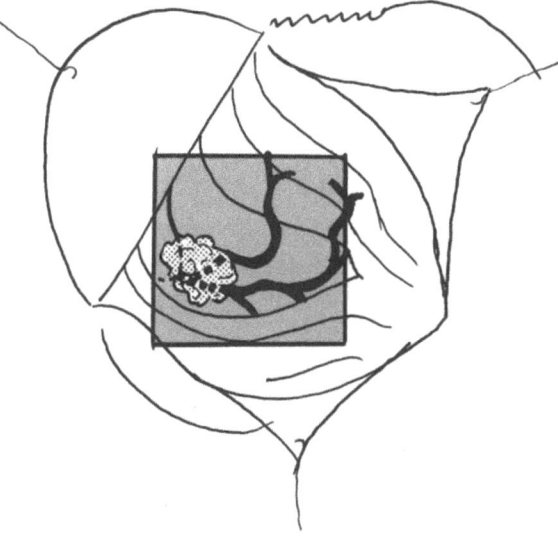

Fig. 19 e

Fig. 19 f

many years. This is of some importance as metallic clips may interfere with postoperative CT Scans.

4. *Uncapping the tumor* (Fig. 19 b): In some cases the tumor is seen as soon as the dura is open, and the capsule may be immediately coagulated. In most patients it is necessary to open the Sylvian fissure. The temporal lobe is gently retracted: the bridging veins between the tip of the lobe and the dura of the sphenoidal ridge are coagulated and cut. The surgeon must take care

to preserve the Sylvian vein: very often, careful dissection even with the naked eye allows the opening of the fissure without damage. The microscope is useful when this dissection of the Sylvian vein is difficult. Some patients have a large and dilated vein the anterior portion of which is included in the tumor: in this situation careful examination of the venous

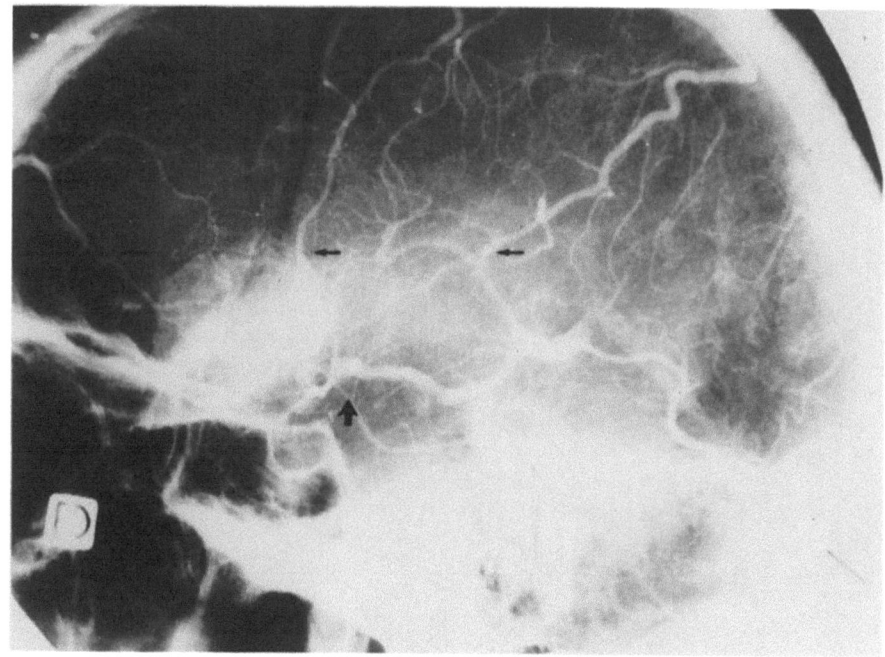

Fig. 20. Venous phase of angiography of a sphenoidal ridge meningioma with stenosis of the Sylvian vein (large arrow) and anastomotic venous drainage towards the superior sagittal sinus (small arrows)

phase of angiography may show a zone of tight stenosis and formation of abnormal anastomotic venous drainage to the superior sagittal sinus or the lateral sinus (Fig. 20) allowing surgical interruption of the anterior part of the vein. Opening the medial aspect of the fissure allows the cerebro-spinal fluid to flow decreasing brain tension. When the tumor is small direct access to the prechiasmatic cistern is easy and early identification of the internal carotid and Sylvian arteries possible. With bigger tumors they are hidden and cannot be identified before removing a part of the tumor.

The next step is appraisal of the tumor: is it of a soft or hard consistency? Is it vascular, and can temporary haemostasis be easily obtained after the removal of a small piece of tumoral tissue? Is it easy to dissect without damage to the surrounding brain tissue? Is it possible to move it? During this time of exploration the surgeon must remain extremely cautious even and especially when the tumor seems to be easy to handle keeping in mind that the danger is out of his sight and control, behind the medial and superior part of the tumor where the internal carotid, Sylvian and particularly the perforating branches must absolutely be preserved. Any heavy traction applied to the capsule, any dissection without direct and good visual control must be avoided. The main risk is to tear small branches of important vessels or to produce brain damage with the risk of creating or increasing epilepsy or functional deficit. When the lateral part of the tumor is not seen at first it is never necessary to locate it by brain needling as after opening the Sylvian fissure, the base of the meningioma can always be found along the sphenoidal ridge. Ventricular tapping is unnecessary as brain tension will decrease after evacuation of cerebrospinal fluid out of the cisterns and removal of a part of the tumor. Lumbar sub-arachnoid drainage may be harzardous with a big tumor. The authors have never had to use it since appropriate preoperative corticoids and neuroleptanalgesia have been routinely employed. A wedge resection of the anterior part of the temporal lobe is seldom necessary, but may be a better solution than hard retraction applied to the brain.

No attempt should be made to dissect the tumor at this stage of the operation. Only gentle retraction may be applied to one or both lips of the Sylvian fissure to obtain a good exposure of the capsule. Large, tortuous tumoral veins must be coagulated. It is easier to apply coagulation with a large forceps because a thin one can produce a hole in the capsule with bleeding difficult to stop. The tumoral capsule is then opened and the removal of the tumor begun.

5. *Removing the outer and central part of the tumor* (Fig. 19 c). If the tumor is soft it may be quickly sucked away. If it is not vascular this method allows the removal of the bulk of the meningioma in a few minutes. The only danger comes from the use of overstrong suction as this may tear the capsule and produce lesions in the brain, its feeding vessels or even the main vascular trunks. Caution must be taken to avoid aspiration into the sucker of vessels running in depressions of the capsule. When the tumor is more vascular, the surgeons' experience is of the utmost importance as it is necessary to understand immediately if the loss of blood, which may be considerable, will be easy to control. Frequently applying cottonoids with gentle pressure and coagulating the main bleeding points will allow the surgeon to proceed in the same way. If bleeding is not controlled easily in this way it is necessary to adopt a slower pace, removing small portions of the tumor with immediate haemostasis. This may be very tedious and time consuming but the neurosurgeon must never attempt a rapid removal of tumor without being absolutely sure of being able to control the bleeding.

With a tumor of hard consistency it is impossible to remove it with suction. The use of curettes allows the surgeon to gut the tumor when it is not too fibrous. Sharp curettes must be used but there is a danger of tearing small vessels if strength is applied to remove more fibrous parts. This frequently happens near the insertion and in the medial part of the meningioma, next to the carotid artery. It is impossible in such a situation to use curettes. Coagulating loops have the same disadvantage so that the authors prefer to exert gentle traction with a forceps and cut a wedge of tumor with scissors. Visual control is necessary at each step. The more fibrous, haemorragic and medial is the part of the tumor being removed, the slower will be the progress of removal.

This stage of removal of the tumor has been greatly simplified by ultrasonic aspiration, but this method, though extremely efficient for the soft parts of the tumor which are not very fibrous, loses a great deal of its utility when dealing with extremely fibrous or calcified meningiomas. In these conditions it works too slowly even when used at its maximum power. It is important to modulate the strength of vibration of the aspirator. When dealing with a soft tumor the consistency of the wall of the vessels is firmer than that of the tumor so that dissection can be performed without danger. Things are quite different with fibrous tumors for which maximum strength of vibration is necessary, involving a risk of lesions of blood vessels and peritumoral brain tissue if the tumor capsule is transgressed.

6. *Removal of the capsule* (Figs. 19 d, e, f). When using the method of tumor gutting described above, particularly with ultransonic resection, the capsule tends to retract inwards and to separate from adjacent structures. By smooth traction with a forceps it can be retracted from the brain. Small vessels are stretched, veins coming out of the tumor are divided, large veins of the cortex dissected and preserved to prevent venous infarction. In the same way small arteries may be interrupted if grooved into the tumor, larger ones must be preserved. The best way is to start posteriorly and medially when the internal portion of the tumor is still left in place. Unless the tumor is very small and the carotid artery and optic nerve identified at an earlier stage, it is wise to identify peripheral lateral branches of the middle cerebral artery and to dissect them starting at their external part and progressing medially (Fig. 19 d) and in a postero-anterior direction by retrograde dissection[10, 12]. In many cases, they are protected by a layer of thickened, opaque arachnoid and the tumor can be totally removed without any risk. In other cases careful dissection is necessary and the use of the microscope and bipolar coagulation necessary. In this situation the dissection is performed with a small spatula or a blunt hook introduced with extreme caution between the vessels and the capsule which is removed millimeter by millimeter. Total removal of tumoral tissue is often possible in this situation, but a long dissection, however carefully it may have been carried out may cause spasm and secondary occlusion of the vessel. It may be wiser

to leave a small piece of tumor than to take the risk of permanent neurological impairment. Some of these meningiomas are unmanageable and their total removal impossible when the carotid artery and the optic nerve are completely surrounded and invaded by the tumor. The author has tried many years ago to remove this kind of tumor with constant failure. Operating on these patients must be thoroughly thought out as these large medial tumors involving the internal carotid artery, and sometimes the bifurcation, the medial and anterior cerebral artery, have an extremely large vascular zone of insertion supplied by meningeal arteries coming from the internal carotid. Stopping the operation when the tumor has been only partially resected may be quite difficult; preventive embolization is impossible and such tumors mark one of the limits of surgical treatment today.

Some authors have proposed preventive extra-intracranial anastomosis in case of lesions of major vessels occurring or anticipated during the operation. The authors of this paper have no personal experience with this procedure. The laser (Nd:Yag or CO_2) has been recommended by some neurosurgeons. A wider experience with this device remains necessary to afford a proper assessment of its utility.

7. *Control of the insertion of the tumor.* The outer part of these deep-seated meningiomas is generally easily controlled, invaded dura is resected and bone thickening removed with rongeurs. The medial part is more difficult to manage. Total removal of the dura and bone involved at this site may produce functional deficit, particularly with involvement of the optic nerve and the carotid artery. As these tumors grow very slowly it may be wiser to coagulate the insertion than to obtain a total resection at the cost of permanent impairment. If the optic nerve is compressed by bony thickening in the foramen, decompression must be made. Resection of tumor bulging into the foramen may be done taking great care to preserve the optic nerve and ophthalmic artery. Sometimes, the dural insertion extends to the roof or the lateral wall of the cavernous sinus. Attempts to remove the insertion totally when the patient has no occulomotor symptoms, should not be considered. The roof and wall may be coagulated. Should any venous bleeding occur it is easy to control even if profuse, by pressing oxycel on the dural defect.

8. *Dural closure.* Once haemostasis of the cavity has been correctly checked and no bleeding point remains—even the smallest one (saline injected in the cavity must remain perfectly clear) dural closure must be considered. If the brain is slack the dura is tightly closed with continuous suture or separate stiches. Stay sutures are placed around the bone flap. A central one is placed to reduce the free space between the dura and the bone flap.

Haemostasis of superficial layers is checked: dura, bone, muscles.

Epidural drainage may be left and will be removed 24 hours later. The bone flap is replaced and sutured and the skin closed in two layers.

Should the brain still be bulging at the end of the operation after removal of the tumor, it may be necessary to leave the dura open to avoid compression, the dural flap is reflected to protect the brain and the defect remaining closed with Oxycel. No drainage must be left in this case.

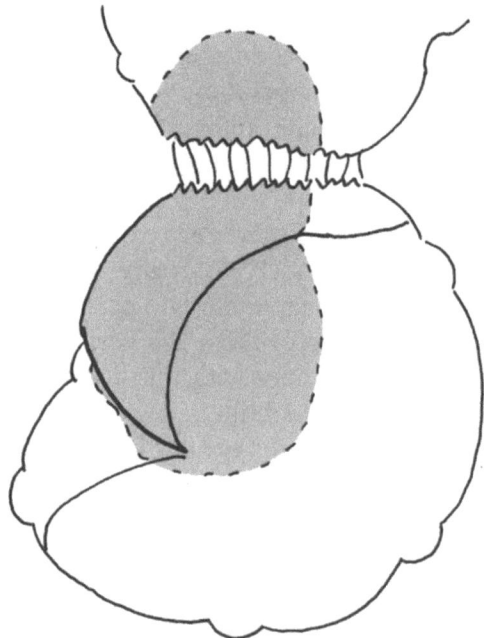

Fig. 21. Resection of lateral pterional meningioma. Grey area: limit of invaded bone and dura to be resected

Removal of Lateral Sphenoidal Wing Meningiomas

Position on the operation table, skin incision and craniotomy are the same as for medial meningiomas. As the tumor insertion is just under the bone flap and the feeding vessels come exclusively from the external carotid artery some particular measures must be taken.

1. The vascularity may be very considerable and preoperative embolization needs to be discussed to reduce bleeding when the bone flap is cut. Bone wax must be used in each burr-hole and in the places where the bone is cut with the Gigli saw or removed with rongeur.

2. The dural insertion of the tumor is just under the bone when adhesion between bone and dura is too tight, the flap must not be turned as dural tearing would occur and a lesion to the brain be produced with subsequent

major neurological disability, especially on the dominant side. The zone of adhesion or of bone invasion must be progressively removed with rongeurs (Fig. 21). All the abnormal bone must be removed to prevent recurrence. Haemostasis is obtained by coagulation and application of bone wax.

3. The dura is opened around the tumoral insertion. When the tumor extends far medially the inner part of the dura is not opened at this stage. By

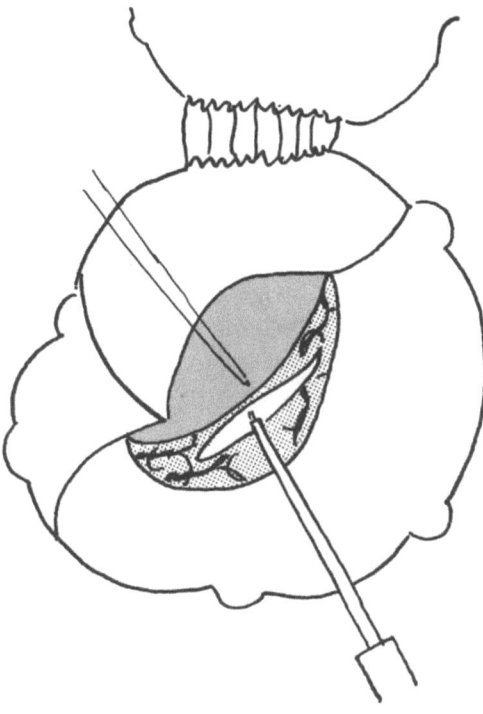

Fig. 22. Pterional meningioma. Dural opening, exposure and capsule incision

careful traction at the edge of the dura near the meningioma the capsule is exposed (Fig. 22) and the resection performed as indicated for medial tumors; coagulation of the capsule, tumor gutting with the same precautions, and careful removal of the capsule. Care must be taken to avoid damage to the middle cerebral artery including the small perforating vessels. The internal carotid artery and optic nerve seldom have connections with the anteromedial part of the tumor. When the abnormal dura has not been completely removed at the beginning of the operation, the opening is completed without risk during this stage as intra-dural assessment of the tumor position becomes possible.

4. As a large part of invaded dura has often been removed it becomes necessary to use a graft for dural closure. The authors favor an epicranial

graft because it may be easily taken during operation and is more pliable than fascia-lata. If the brain is slack, closure of bone and skin is made as described before. When the brain is still bulging after the removal of the tumor the dura may be left open. A large piece of oxycel is used to close the defect and protect the brain. It will quickly give rise to a neodura. In this case the bone flap is replaced and loosely fixed. No drainage is used as it

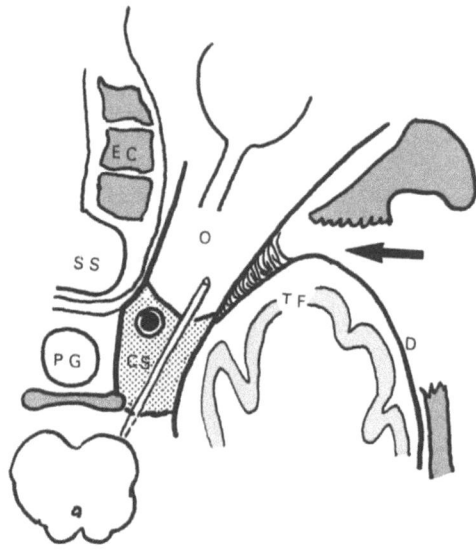

Fig. 23. Schematic drawing of relation of bone, dura and external wall of cavernous sinus. The black arrow shows the key point to separate temporal and cavernous sinus dura. *CS*: cavernous sinus, *D*: dura, *EC*: ethnoidal cell, *O*: optic nerve, *PG*: pituitary gland, *SS*: sphenoidal sinus. (From the anatomic atlas by Delmas and Pertuiset)

could be the origin of a cerebro-spinal fluid leak. The collection of cerebrospinal fluid which may appear under the skin generally diminishes in some days. The resorption may be helped by repeated lumbar punctures evacuating fluid until the bulging under the skin disappears, and by the use of a diuretic drug (Acetazolamide).

Management of "en plaque" and Invasive Meningiomas

These tumors extend throughout the temporo-spheno-orbital region through the temporal bone (sometimes invading the temporal muscle) spreading into the orbit the temporal fossa. They may cross the midline to invade a large part of the base of the skull. Neurosurgeons have devised operations to try to offer a complete, or as complete as possible, resection of these tumors. Bonnal[2, 3], Guiot and Derome[1, 14] have published results of large series and made a detailed description of this complex procedure. Only the general principles of the resection will be given here. Removing all the sphenoidal ridge including a part of the anterior clinoid process, the floor of the temporal fossa the intra-orbital part of the tumor with resection of the roof of the orbit, the invaded periorbitum, the lateral wall of the orbit is a major operation and necessitates careful reconstruction of the orbit and of the base of the skull. The dural adhesion with the bone is easy to separate at the lateral aspect of the temporal fossa but is very firm at the sphenoidal ridge and the inner part of the temporal floor. The key to this dissection (Pertuiset[6]) along the lateral wall of the cavernous sinus is given by cutting the adhesion between the temporal dura and the lateral part of the supra-orbital fissure. This dissection gives access to the foramen rotundum, the foramen ovale and foramen spinosum (Fig. 23). Resection of the bone surrounding the superior and lateral parts of the supraorbital fissure is also possible. Resecting the anterior clinoid process must be very carefully done and it is not always possible to perform total resection. Resection of thick, very hard bone is made with rongeurs and power drills.

Such an extensive resection needs careful reconstruction with bone grafts. Details of this plastic stage of the operation is described by Derome[7] and Lesoin[11].

Postoperative Care

After operation the patients are placed in the intensive care unit where proper monitoring and nursing is available. The duration of artificial ventilation must be discussed in each case. It may be helpful to maintain it for several days when difficulties have arisen during operation and when the brain was not perfectly slack at the end of the procedure.

Corticoids are continued for a few days after operation, unless they have not been given before operation, and when brain swelling has not been present. Antibiotics are given routinely for five days. Systematic administration of anitepileptic treatment is discussed, it is mandatory when epileptic seizures have occurred before.

Scalp sutures are removed on the seventh day after the operation.

Results

The results after operation in the series presented here are shown in Table 3. It is quite clear that the removal of sphenoidal ridge meningiomas is a serious operation with many hazards. It must be stressed that these patients have been operated on over a long period of time and that the results have been better in later years owing to progress in surgical technique and anaesthesia. The choice of the patients is also most important. When

Table 3. *Results of 44 Operated Sphenoidal Ridge Meningiomas* (percentage)

	All tumors	Medial	Lateral
Mortality	23	27	17
Severely impaired	14	23	12
Normal life or mild impairment	63	50	71

looking at the results of resection of lateral meningioma, some of our patients who died of general infection or cardio-respiratory failure were poor candidates for major surgery and would not be operated today as the troubles they had were not life-threatening nor severely disabling. With these precautions, lateral, globular meningiomas of small or medium size occuring in patients in good health can reasonably be removed by every well-trained neurosurgeon with good results. Operating on a medial, invasive or "en plaque" meningioma is quite another matter. In the first case the danger comes from the relation of the tumor with vital structures particularly the carotid and middle cerebral arteries and the risk of major bleeding, and proximity of the optic nerve, orbit and cavernous sinus. In 4 of our 9 patients who died or suffered major hemiplegia, arterial obstruction of major vessels was found at angiography or pathologic study. This risk has always be stressed in large series: (Pertuiset, Guyot[9] and Fischer[8]). Other complications are aggravation of visual impairment or diplopia, and post-operative hematoma. Such dangerous operations must be executed by surgeons with great experience with this region and well trained in microsurgery. This is also true for the treatment of invasive and "en plaque" meningiomas as the large resections proposed now necessitate an anatomical and pathological understanding of regions which are not familiar to all neurosurgeons.

The appearance or aggravation of seizures may also occur when lesions are created in the cortex at operation. This emphasizes the care necessary in the dissection of the tumor capsule. The rate of recurrence is one of the highest with intracranial meningiomas: 34% of recurrence or progression at 5 years and 54% at ten years in the series of Mirimanoff[13], for sphenoidal ridge meningiomas but only of 3% and 25% at the convexity. For Philippon[16] 25% for medial and 9–7% for lateral meningiomas at 10 years.

These last results are in accordance with the author's experience.

When total removal has not been possible the question of supplementing the operation by radiotherapy must be raised, as the growth of meningiomas has been reported to be decreased in this way. The recent discovery of receptors to sex-steroid hormones in meningiomas[17] has not resulted in clinical application as yet but may be promising in the future to control the development of these tumors.

Acknowledgments

The authors thank Prof. B. Pertuiset who authorized the addition of his personal cases to the series and Prof. J. Metzger who carried out the neuroradiological investigations. Mrs. C. Callec and Miss G. Coquillette must also be thanked for their technical assistance.

References

1. Basso A, Carrizo A *et al.* (1978) La chirurgie des tumeurs sphéno-orbitaires. Neurochirurgie 24: 71–82
2. Bonnal A, Castermans A *et al.* (1972) Les méningiomes des étages antérieurs et moyen de la base du crâne. Conduite à tenir vis à vis des envahissements osseux et des prolongements dans les cavités de la face. Neurochirurgie 18: 441–451
3. Bonnal J, Thibaut A *et al.* (1980) Invading meningiomas of the sphenoid ridge. J Neurosurg 53: 587–599
4. Cook NA (1971) Total removal of large global meningiomas at the medial aspect of the sphenoid ridge. J Neurosurg 34: 112–113
5. Cushing H, Eisenhardt L (1962) Meningiomas. Their classification, regional behaviour, life history and surgical results. Hafner, New York, pp 298–387
6. Delmas A, Pertuiset B (1959) Topométrie crânio-encéphalique chez l'home. Masson, Paris, Ch C Thomas, Springfield Ill
7. Derome P *et al.* (1972) Les tumeurs sphéno-orbitaires. Possibilités d'exérèse et de réparation chirurgicales. Neurochirurgie 18 [Suppl] 1
8. Fischer G, Fischer C, Mansuy L (1973) Pronostic chirurgical des méningiomes de l'arête sphénoïdale. Neurochirurgie 19: 323–346
9. Guyot JF, Vouyouklakis D, Pertuiset B (1967) Méningiomes de l'arête sphénoïdale à propos de 50 cas. Neurochirurgie 13: 571–584
10. Kempe GL (1968) Operative neurosurgery, vol 1. Springer, Berlin Heidelberg New York, pp 109–118
11. Lesoin F, Pellerin P *et al.* (1986) Intérêt de la voie d'abord orbito-fronto-temporo-malaire dans l'exérèse de certains méningiomes de l'arête sphénoïdale. Neurochirurgie 32: 154–160
12. Mac Carthy CS Meningiomas of the sphenoidal ridge. J Neurosurg 36: 114–120
13. Mirimanoff RO, Dosoretz DE *et al.* (1985) Meningioma analysis of recurrence and progression following neurosurgical resection. J Neurosurg 62: 18–24
14. Pompili A, Derome J *et al.* (1982) Hyperostosing meningiomas of the sphenoidal ridge. Clinical feature surgical therapy and long term observations. Review of 49 cases. Surg Neurol 17: 411–416
15. Pertuiset B, Farah S, Clayes L *et al.* (1985) Operability of intracranial meningiomas. Personal series of 353 cases. Acta Neurochir (Wien) 76: 2–11

16. Philippon *et al.* (1986) Les méningiomas récidivants. Neurochirurgie 32 [Suppl] 1
17. Poisson M, Magdelenat H, Foncin JF *et al.* (1980) Récepteurs d'ostrogènes et de progesterone dans les méningiomes. Etude de 22 cas. Rev Neurol (Paris) 136: 193–203
18. Teasdale E, Patterson J *et al.* (1984) Subselective preoperative embolization for meningiomas. J Neurosurg 60: 506–514
19. Watts C (1985) Sphenoid wing meningioma. In: Long DM (ed) Current therapy in neurological surgery. Decker, Toronto, pp 14–16

Congenital Spinal Cord Tumors in Children

H. J. Hoffman, R. W. Griebel, and E. B. Hendrick

Division of Neurosurgery, University of Toronto and The Hospital for Sick Children, Toronto, Ontario (Canada)

With 7 Figures

Contents

1. Introduction: Comments on Classification and Embryology 176
2. Dermoid and Epidermoid Cysts 177
   A. Embryology 177
   B. Clinical Features 178
   C. Radiology 182
   D. Surgical Management 183
   E. Discussion 185
3. Neurenteric Cysts 185
   A. Embryology 185
   B. Clinical Features 186
   C. Radiology 187
   D. Surgical Management 187
   E. Discussion 188
4. Teratomatous Cysts 188
   A. Embryology 188
   B. Clinical Features 189
   C. Case Presentation 189
5. Intraspinal Teratomas 191
   A. Embryology 191
   B. Clinical Features 191
   C. Case Presentations 191
6. Lipomas 192
   A. Embryology 192
   B. Clinical Features 193
   C. Case Presentations 193

7. Spinal Arachnoid Cysts 195
8. Congenital Malignant Tumors of the Spine 196
9. Summary 197
References 197

1. Introduction

Congenital spinal tumors may broadly be defined as spinal tumors present at birth. The term conventionally refers to a pathologically diverse group of lesions including dermoid and epidermoid cysts, neurenteric cysts, teratomatous cysts and teratomas, lipomas and arachnoid cysts. Although individual types of congenital tumors are relatively rare, as a group they constitute 18–30% of all paediatric spinal tumors [10, 11, 22]. By comparison this same group comprises only 2–3% of all adult spinal neoplasms [10].

This paper reviews the embryology, clinical features, radiology and surgical management of congenital spinal tumors in the paediatric population. It is based on the experience of the last 20 years at the Hospital for Sick Children in Toronto. The various oncotypes will be discussed independently.

As many congenital tumors inhabit the boundary between malformation and neoplasm, problems with categorization arise. Venous angiomas and spinal arteriovenous malformations, for example, are primarily malformations and are not included in this review. Lipomyelomeningocoeles are likewise primarily malformations. The growth of these fatty tissue masses is not autonomous, and their clinical relevance is usually the result of cord tethering rather than mass effect. Certain lipomas, however, do behave as bonafide neoplasmas, and these will be considered as congenital tumors.

The term congenital also applies to malignant neoplasms presenting within two to six months of birth and which presumably arise in utero [53, 58]. Although primary congenital neurectodermal tumors of the brain are not uncommon, these have not been reported to occur in the spinal cord. Neuroblastomas, however, may present with compression of the cord at birth or shortly thereafter, and will be discussed.

The dysgenetic syndromes, or phakomatoses, are associated with a variety of CNS tumors which may involve the spinal cord including neurofibromas, meningiomas and hemangioblastomas. Although there is a hereditary predisposition to tumor formation in affected patients, the tumors themselves arise from cellular metaplasia later in life. These lesions are therefore not truly congenital and are not reviewed.

Embryological consideration of benign congenital tumors must take into account the intimate association between these lesions and spina bifida.

Dermoid cysts and sinuses, lipomas, neurenteric cysts, arachnoid cysts and teratomatous cysts have all been reported in conjunction with varying degrees of spinal cord and vertebral dysplasia. This association is dramatically exemplified in a case report by Wakai and Chiu were a young girl presented with a combination of spina bifida occulta, lipoma, congenital dermal sinus and arachnoid cyst[59]. All of these congenital masses may occur as isolated findings however, so a strict causal relationship between spina bifida and the tumors cannot be established. Moreover, most cases of spina bifida are unassociated with spinal tumors.

It is known that the developmental "accidents" that result in both spina bifida and the congenital tumors occur early in the embryonic period when the basic cell layers differentiate and primitive organ systems are formed[17]. Many authors have attempted to develop a unified theory of dysraphism which would explain the full spectrum of dysraphism and the associated lesions including congenital tumors. Incongruities persist however, and to date no all encompassing explanation satisfactorily accomplishes this task. We shall briefly review therefore, a selection of proposed mechanisms of tumor formation as the various tumor types are presented.

2. Dermoid and Epidermoid Cysts

A. Embryology

Boestroem in 1897 attributed the development of spinal dermoid tumors to an error in neural tube closure with an inclusion of embryonal epidermal ectoderm[4]. This concept was reiterated by Walker and Bucy and is still widely accepted[61]. It is postulated that, for undefined reasons, an adhesion develops between the cutaneous ectoderm and the neurectoderm, thus preventing the usual clean separation of these layers. The collection of cells remaining adherent to, or embedded in, the closed neural tube engenders subsequent cyst formation. An epidermal tract almost invariably persists between the neurectodermal inclusion and the epidermis, which provides a potential conduit for surface organisms to the subarachnoid space. The tract itself may interfere with mesodermal condensation resulting in a posterior spina bifida.

Padget placed a variation of this theory in wider perspective[43, 44]. She maintains that neuroschisis, or clefting of the closed neural tube in the presomatic embryo is a relatively common occurence which may or may not heal in the course of ongoing embryonic development. An unhealed schisis presents as a myelomeningocoele, or variant thereof. Following schisis, in some instances, a degree of secondary continuity is established between the damaged neural and cutaneous ectoderm. The intervening mesoderm is damaged by a fluid bleb which accumulates at the site. The dermoid sinus and bifid vertebrae thus remain as the "scars" of a healed neuroschisis. Seen in this perspective it may be anticipated that ectodermal inclusions would also be found in association with spina bifida aperta and cystica.

B. Clinical Features

Over a 20 year period (1965–1985) 29 children with dermoid or epidermoid spinal cysts were treated at the Hospital for Sick Children. By comparison over a similar period, 13 children with intracranial dermoid or epidermoid tumors were operated upon and 12 children underwent excision of dermal sinuses without an intraspinal cystic component.

Fourteen of the children with spinal dermoid and epidermoid tumors had an associated myelomeningocoele or meningocoele. Because of differences in presentation, radiological assessment and surgical results between these children and the group without obvious dysraphism the two groups will be discussed separately.

Dermoid and epidermoid cysts may be either acquired or congenital lesions. Both clinical and experimental evidence supports the concept that skin fragments implanted by lumbar puncture needles may provide a nidus for future cyst development[5, 41, 56]. Three of the 15 children without myelomeningocoeles in this series had lumbar cysts and a history of having had lumbar punctures in infancy. As none of these children had associated vertebral or cutaneous anomalies it seems probable that these cases were of iatrogenic origin.

Bryant and Dayan first reported an inclusion dermoid in a patient with a treated myelocystocoele and concluded it resulted from a fragment of epidermis enclosed during the surgical repair[7]. Not all dermoid cysts in myelomeningocoele patients, however, result from surgical implantations. One of the cysts in our patients was found at the time of initial closure, 3 patients had cysts remote from the repaired dysraphic defect and 2 patients has associated dermal sinus tracts. While these 6 cases were certainly of congenital origin, the aetiology of the other 8 remains speculative.

Histological analysis determined all of the cysts in the myelomeningocoele patients to be dermoids. All but 4 of the patients in the non-myelomeningocoele group also had dermoid cysts. Three of these exceptions were the patients who had a history of lumbar punctures as infants and all 3 of these had epidermoid cysts.

The age, sex, presenting symptoms and associated deformities of these two groups of children are outlined in Tables 1 and 2. It will be noted that the sex ratios between the two groups are reversed; 9 males: 6 females in the nonmyelomeningocoele group and 3 males: 11 females in the myelomeningocoele group.

Table 1. *Clinical Features of Dermoid and Epidermoid Cysts in Patients Not Having a Myelomeningocele.* All patients with epidermoid cysts except one, had a history of lumbar puncture as infants

Patient	Sex	Age at surgery	Presenting symptom	Dermal stigmata	Vertebral anomalies	Level of lesion
P.A.	m	2 ½ yrs	back pain, monoplegia	sinus tract	none	L_1–L_4
C.W.	f	1 month	skin anomaly	naevus and sinus	none	T_{12}–L_1
H.B.	f	14 yrs	meningitis	sinus tract	bifid T_2	T_1–T_2
J.S.	m	9 mos	meningitis	sinus tract	bifid T_7–T_8	T_7
P.L.	m	1 yr	meningitis	sinus tract	hypoplastic spinous process	T_{12}–L_3
P.T.	m	7 mos	meningitis	naevus and sinus tract	bifid L_5–S_1	L_1–L_2
C.G.	m	11 yrs	weak L leg	dermal sinus	bifid T_5	T_5
M.H.	m	3 ½ yrs	meningitis	dermal sinus	none	S_1–S_3
L.S.	f	14 mos	meningitis	naevus and dermal sinus	bifid S_1	L_2–L_3
G.D.	m	3 mos	dermal sinus	dermal sinus	bifid T_3–T_4	T_{2-3}
D.L.	m	5 1/2 yrs	leg pain	none	interpedicular widening	T_{12}–L_1
Epidermoids:						
V.F.	f	6 ½ yrs	back pain	none	none	L_2–L_4
G.J.	m	12 yrs	back pain	none	none	L_5–S_1
S.L.	f	8 yrs	calf pain	none	none	L_2–L_3
U.F.	f	17 yrs	back pain, leg weakness	none	klippel-feil, spina bifida occulta of entire spine	L_1–L_2

Table 2. Clinical Features of Patients with Dermoid Cysts and Spina bifida aperta

Patient	Sex	Age at surgery	Presenting symptom	Level of lesion	Associated anomalies
P.H.	F	12 yrs	Progressive spasticity and enuresis	L_4–L_5	sacral meningocoele
I.W.	f	8 yrs	deteriorating gait, increased spasticity	L_1	lumbar myelomeningocoele, diastematomyelia L_3
G.E.	m	4 yrs	deteriorating gait	L_2–L_3	meningocoele, leg pain tethered cord
S.C.	f	6 yrs	back pain	T_4	myelomeningocoele, diastematomyelia
D.C.	f	4 yrs	deteriorating gait	L_2	myelomeningocoele diastematomyelia
B.D.	m	13 yrs	progressive spasticity	L_3–L_4	myelomeningocoele
D.W.	f	17 yrs	deteriorating gait	T_{11}–T_{12}	myelomeningocoele diastematomyelia
S.W.	f	3 ½ yrs	urinary and fecal incontinence	S_1	myelomeningocoele, diastematomyelia
W.C.	m	8 mos	discharging sinus tract	L_5–S_1	myelomeningocoele
L.A.	f	5 yrs	progressive spasticity	T_{11}–T_{12}	meningocoele
T.H.	f	10 yrs	deteriorating gait	T_{12}–L_2	sacral myelomeningocoele
C.E.	f	2 weeks	apneic spells	C_1	lumbar myelomeningocoele
C.P.	f	11 yrs	increasing spasticity	T_{12}–L_1	myelomeningocoele
B.B.	f	1 month	incidental finding	L_3	lumbar myelomeningocoele

Age at time of surgery for the myelodysplastic group ranged from 1 month to 17 years (M = 4.2 years) with five of these patients less than a year of age at time of surgery. The age range for the myelomeningocoele patients was from 2 weeks to 17 years (M = 7.3 years).

The tumors in both groups of children were situated predominantly in

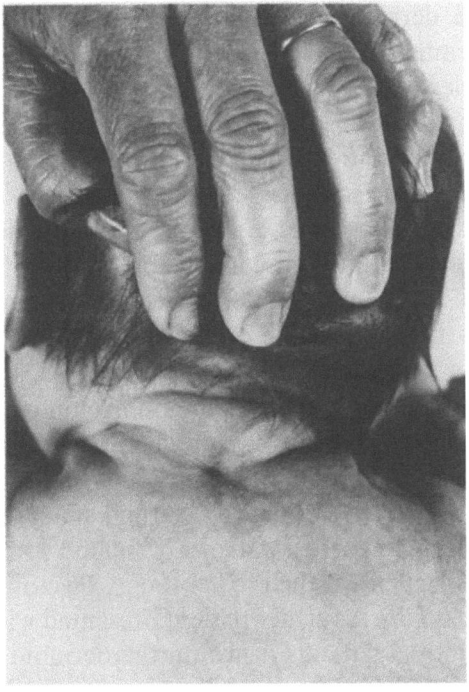

Fig. 1. Midline dermal pit associated with an underlying sinus tract and dermoid tumor

the lumbar region. Eleven of the 14 dermoids in the myelodysplastic group were encountered at the level of the dysraphic malformation. In the other 3 patients the tumors were situated cephalad to the spinal defect.

Three of the cysts in the nonmyelomeningocoele group were enclosed within the cord whereas 6 of the myelomeningocoele group were mainly intramedullary. Three cysts in all transgressed the dural boundary and one cyst was found lying completely extradurally. In 3 children multiple cysts were found.

A dermal sinus tract was present in 10 of the 11 children with histologically proven dermoids and no myelomeningocoele. The four children with epidermoids had no sinus tract or other midline cutaneous stigmata. In 6 cases the opening of the dermal sinus was described as a simple dermal pit (Fig. 1) while in the other 5 it was associated with port wine stains or hairy tufts. Anomalies in other body systems or associated CNS anomalies were not found apart from one child who had polydactly and syndactly in addition to his spinal lesion.

A dermal sinus was seen in only 2 of the 14 children with myelomeningocoeles. In one child a discharging infected dermal sinus was apparent at the lower end of a lumbar myelomeningocoele repair. The other child had a cervical sinus and dermoid in association with a lumbar myelomeningocoele. It is of interest that five of the patients with dermoids and myelomeningocoeles also had diastematomyelia.

The presenting features of the two groups were quite distinct. None of the myelomeningocoele patients presented with CNS infections whereas 6 of their counterparts suffered 12 separate episodes of meningitis and 2 patients had spinal cord abscesses. Five of these six children were infants at the time of infection. This difference in infection rates between the groups is undoubtedly the result of the dermal fistulae in the nonmyelomeningocoele group. E. coli was the responsible organism in the majority of cases but proteus, citrobacter and listeria were also isolated. Back pain was the most common presenting symptom in the older nonmyelomeningocoele children. Four children presented with back pain whereas 2 others presented with lower limb weakness and gait difficulties. Three infants underwent prophylactic surgery upon recognition of a dermal sinus following birth.

Ten of the myelomeningocoele patients presented with deteriorating gait and/or increasing spasticity of their lower limbs. Two of these patients also had significant back pain. A further patient presented with enuresis. In this group there were only 4 patients in whom the dermoid cyst was the only significant pathology, the remainder having diastematomyelia or tethered cords in addition to the cysts. One infant had his dermoid encountered upon resection of a draining dermal sinus, and in another the dermoid was found incidentally at the time of initial myelomeningocoele repair. A 2 week old infant with a cervical dermoid presented with apneic spells and regurgitation of feeds.

C. Radiological Investigations

Myelography was performed in all patients but one of the nonmyelomeningocoele group and the mass lesion was demonstrated in all these cases. A CT metrizamide scan of the spine was carried out in 7 cases and invariably outlined a mass of low attenuation values (Fig. 2).

Eleven of the myelodysplastic patients underwent myelography. In only 6 of these were the dermoids recognized as mass lesions. In the other 5 cases the tumors were small nodules of less than 1 cm. CT scanning was carried out in 7 cases, and small dermoids were not demonstrated in 4 of these.

D. Surgical Management and Follow-up

Treatment of the cysts was surgical in all cases. In 9 of the patients with myelomeningocoeles associated pathology was the prime indication for surgical intervention and the dermoid inclusions were encountered incidentally and removed. In 5 of this group of patients however, the dermoid tumor was the significant pathology and all of these patients showed improvement

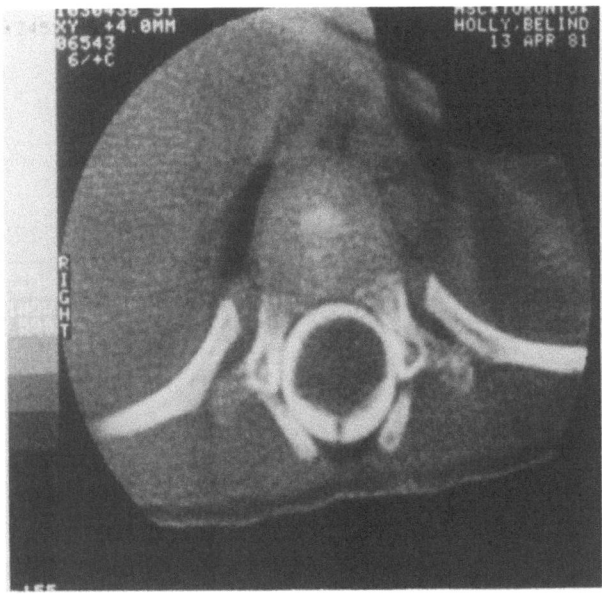

Fig. 2. CT metrizamide scan demonstrating an intramedullary dermoid cyst. Not the communicating sinus tract transversing the subarachnoid space posteriorly

of their symptoms following surgery. Follow-up of this group ranged from 1 year to 13 years (mean 4.3 years). There were no symptomatic cyst recurrences over this period. One of the children died of meningitis 2 years following dermoid excision, but it was not established if this related in any way to his spinal pathology.

All of the patients without myelomeningocoeles underwent excision of their dermal sinus tracts, laminectomies and intradural removal of the tumors (Fig. 3). Six of the 16 patients required repeat surgical procedures. In 3 infants infections within months of the initial surgery prompted re-exploration for removal of residual tumor. Two children developed meningitis 1 and 4 months postoperatively and the third developed a spinal cord abscess 2 months after the initial operation. In all cases the dermal sinus had been excised indicating that contaminating bacteria resided in the residual cyst. Re-excision of residual tumor resulted in a favorable outcome in all 3 cases.

In three other cases an initial partial resection resulted in tumor recurrence presenting with cord compression. One child underwent 4 repeat procedures between 2 and 10 years after the first operation. On each occasion recurrent tumor was encountered and excised. More recently, he developed a syrinx at the site of tumor excision which required shunting.

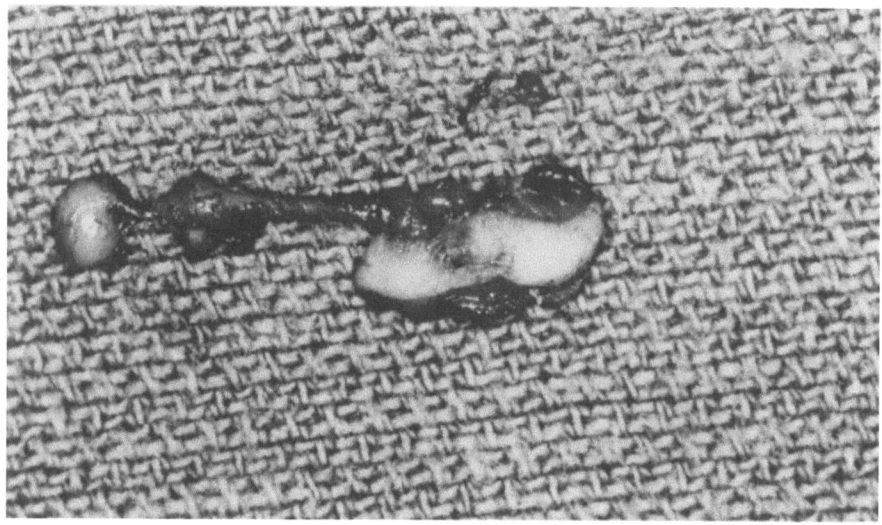

Fig. 3. Surgical specimen following excision of dermal sinus pit and the dermal sinus tract terminating in a small dermoid cyst

Inspite of these 6 surgical procedures he is only left with moderate weakness of ankle dorsiflexion.

A 17-year-old girl with a cystic epidermoid of the conus had a partial excision of the cyst wall which required re-exploration and re-drainage of the cyst 7 months after her first procedure. A third exploration was performed when leg power again deteriorated at which time a silastic catheter was situated in the cyst and connected to an Ommaya reservoir. Thereafter, recurrent symptoms were relieved by reservoir aspiration[25].

A third child had tumor recurrence and surgery 2 years after an incomplete excision and was thereafter asymptomatic. Five years later, however, he was killed in a tobaganning accident and at autopsy was found to again have a significant (1.25 × 0.8 cm) recurrence of his dermoid.

Follow-up of the nonmyelomeningocoele group ranged from 2 months to 16 years (mean 3.5 years). Two of the patients had residual bowel or bladder dysfunction, 1 suffered weakness and sensory impairment in his left leg, and 1 patient who presented with a spinal cord abscess suffered a permanent weakness of intrinsic hand muscles. There was no operative mortality in either group, one post-operative wound infection and one post-operative pseudomeningocoele.

E. Discussion

A dermal pit above the sacrococcygeal area serves as a warning sign that the child may have a sinus tract extending into the spinal canal. Further investigations (metrizamide CT scan or MRI scan) are indicated prior to excision of the sinus.

Sacroccygeal dimples may be tethered caudally to the coccyx, in which case they are benign, or rostrally, in which case they may have an intraspinal extension. This situation can be differentiated by applying traction to the sinus by fingers placed on either side of the defect. If downward traction of the sinus causes the ostium to gape this indicates that tethering is caudal and further investigation is not indicated. Narrowing and puckering of the ostium on downward stretch however indicates rostral tethering and further radiological studies should be carried out.

Partial excision of intradural dermoid leaves the child at risk both for further infections and recurrence of the mass. With modern microsurgical techniques complete cyst wall excision should be possible at the initial procedure.

Dermoid tumors in the myelomeningocoele population are most often a coincidental finding. They may, however, be responsible for functional deterioration in these patients and must be considered in the differential diagnosis of increasing lower limb spasticity and back pain in these children.

3. Neurenteric Cysts

A. Embryology

The pathoembryology of neurenteric cysts affects all three germ layers. Several different theories have been proposed to explain the anomalies encountered with these cysts, but they all center on the formation of an abnormal or persistent connection between the primitive gut analog and the neurectoderm.

In normal fetal development a caudal fistula known as the neurenteric canal forms as an transitory communication between the amniotic cavity and the yolk sac at about 3 weeks. Bremer postulated that the persistence of this fistula, or the abnormal development of a similar fistula at a more rostral level, provided the basis for neurenteric cyst formation[6]. The presence of this fistula was felt to inhibit normal notochord formation and hence the associated vertebral anomalies. Bentley and Smith, however, argued that the primary pathology was in fact abnormal notochord formation[3]. They proposed that duplication or separation of the notochord would create a gap permitting endoderm to herniate through and attach to the dorsal ectoderm. Trapped remnants posterior to the notochord would form neurenteric cysts or, with further dorsal extension, an enteric fistula. The split notochord would give rise to a variety of boney malformations including diastematomyelic spurs, block vertebrae, hemivertebrae or butterfly vertebrae. The combination of these anomalies with spina bifida and a myelomeningocele as reported by Odake, would not be inconsistent with this concept[42].

Other authors have proposed a primary adhesion between the endoderm and ectoderm of the primitive bilaminar embryo which prevents interposition of mesoderm and notochord formation. The characteristic vertebral and enteric malformations result from a situation somewhat analogous to dermoid cyst and sinus formation[2, 21, 49]. Padget did not specifically address the various malformations associated with neurenteric cysts in her schema of neuroschisis, but did conceptualize that neuroschisis of the developing cord may occur anteriorly as well as posteriorly.

B. Clinical Features and Presentation

Between 1960 and 1984 14 children with spinal enterogenous cysts were treated at the Hospital for Sick Children. The group included 7 males and 7 females.

Ten of the 14 cysts were located in the upper thoracic (T_1–T_3) region. One, associated with a diastematomyelia and myelomeningocele, was situated at $T_{11/12}$ along with the dysraphic lesion. The 3 remaining cysts were cervical in location.

All cysts were intradural. They were extramedullary in 8 cases while in 6 cases they were either wholly or partially embedded in the cord.

There were several patterns of presentation. The infants tended to present with infectious complications or respiratory distress, while the older children presented with evidence of cord compression. Three children (aged 2 days, 1 year and 4 years) presented with meningitis and 2 of these died. A 6 week old female with respiratory distress had a large dumbell cyst, the intrathoracic component of which caused respiratory embarrassment.

Nine children presented with signs and symptoms of spinal cord compression. In 5 cases cervical or upper thoracic pain was a major symptom, and in one case the only symptom, of the intraspinal lesion.

Three of the children were noted to have a Horner's syndrome at time of initial examination. Seven patients had a kyphoscoliosis, 1 had a marked gibbus deformity. Two of the children had an interscapular hairy patch overlying the cyst and another had an associated myelomeningocoele and diastematomyelia.

C. Radiology

Plain X-rays or tomograms of the spine revealed vertebral body and/or posterior arch abnormalities in 12 of the 14 cases. The boney anomalies depicted radiographically included hemivertebrae (4 cases), vertebral body fusion (5 cases), widened canal (4 cases), canal of Kovalesky (3 cases),

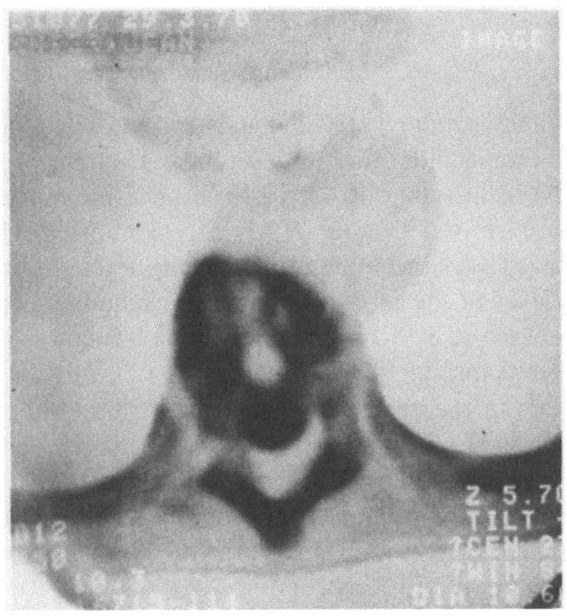

Fig. 4. CT scan of a neurentric cyst demonstrating the canal of Kovalesky

diastematomyelia (3 cases), spina bifida (2 cases) and butterfly vertebra (1 case).

Myelograms were obtained in 13 of the children and CT scans in 4. Five myelograms demonstrated a complete block while an intra or extramedullary mass was outlined in the remainder. CT scanning in addition to demonstrating the mass lesion, also defined the boney anomalies (Fig. 4).

D. Surgical Management and Follow-up

Thirteen of the children underwent surgical excisions of their cysts. An infant who died of meningitis at 3 days of age was not treated surgically.

In 12 cases a posterior approach including laminectomy and intradural exposure of the cyst was used. A partial removal of the cyst wall was achieved in 5 cases and a complete removal of the cyst in 7 cases. In one patient, a partial removal of the cyst was obtained via an anterior approach at the time of removal of a mediastinal mass. Two other children required thoracotomies in addition to laminectomies for excision of an inthoracic component of the lesion.

There was one post-operative death. A 4-year-old boy developed pseudomonas meningitis and pneumonia following his surgery. The other children were followed for 6 months to 3½ years after surgery (mean 1½ years). None of the children demonstrated worsening of their neurological status and there were no recurrences in spite of partial removals in 6 instances.

E. Discussion

In 1978, Holmes et al. reviewed 26 paediatric cases with neurenteric cyst reported in the English literature to that time[27]. He found a predominance of males (18:8) in his compiled series with an average age of 4 years 8 months compared to our average of 4 years 11 months. The cysts in his review also showed a predilection for the upper thoracic and cervical region.

Over half (54%) of the cases in Holmes series had associated vertebral anomalies compared to 86% of our series. Spina bifida was present in 26% of Holmes patients and in 14% of our group.

A spinal mass in the presence of anterior vertebral anomalies should bring the diagnosis of a neurenteric cyst to mind; in the absence of boney anomalies there are no characteristic features which may lead to a diagnosis.

Nineteen percent of the patients in Holmes review died of meningitis or associated anomalies. Our mortality rate of 14% reinforces the lethal potential of meningitis with this condition.

The risk of recurrence following a partial resection of the cyst wall appears slight. With 6 partial resections in our series there have been no recurrences to date. Holmes, however, reported one recurrence following partial resection and the authors are aware of another adult case where a cyst recurred following incomplete excision. Complete excision under magnified vision is therefore advised.

4. Teratomatous Cysts

A. Embryology

Teratomatous cysts may contain, in addition to a cuboidal or columnar epithelium, collagenous fibrous tissue, smooth or striated muscle, mucous or serous glands, cartilage, bone, myelinated fibers, fat cells, paccinian corpuscles, ganglion cells and glial tissue. The origin of these cysts remains controversial. Kubie and Fulton felt they may be either of ependymal origin, arising as a diverticulum of the central canal or possibly an inclusion of fetal epithelium[34]. Initially, teratomatous cysts were felt to contain only ectodermal and mesodermal elements but cases containing endodermal derivatives have been described[8, 28]. Bucy and Buchanan rejected the idea that these tumors arose from inclusions of surface material and theorized they were the result of "maldevelopment of the ovum, the misplacement of some multipotential germinal cells early in embryonic development"[8]. The determination of nuclear sex of the epithelial cells of teratomatous cysts has advanced the concept that these tumors arise from the misplacement of germ cells. Hoefnagel et al. and Rewcastle and Francoeur found 20–30% of cells in the cyst walls of their male patients to be chromatin positive (i.e., female)[23, 48]. This data suggests that haploid gonadal cells displaced to the dorsal midline may fuse and initiate tumor formation. Fusion of two haploid X chromosomes in a male patient would give a chromatin positive cell[50]. This theory is at odds with the concept of the tumor progenitor cells arising directly from the toti potential cells of Hensen's node. Rhaney and Barclay implicated these cells as the probable source of teratomatous cysts[49].

Differentiating teratomatous cysts from neurenteric cysts histologically may be difficult and some authors have indicated that the distinction is more likely to be made on the basis of embryonic origin than microscopic appearances[33]. Both cyst types may be composed of well differentiated epithelium with a basement membrane. Both may contain mucin and other well differentiated cell types. Rosenbaum, however, has pointed out that neurenteric cysts should not contain the ciliated epithelium that usually characterizes the teratomatous cell wall[50]. A nonciliated epithelium resembling intestinal tissue with associated anterior vertebral anomalies will confirm the true nature of these lesions. Palmas recent review, however, indicates that associated vertebral or visceral anomalies are not always found with neurenteric cysts[45].

B. Clinical Features

Although there is a continuum of histological features between the trigerminal teratoma and the ciliated, epithelial lined teratomatous cyst, the histology of the cysts is sufficiently distinct to warrant categorization in a separate subgroup[50]. The entity is very rare. Rosenbaum documented 22 cases of teratomatous cysts in the literature to 1978[50]. Only 10 of the reported cases have been in the pediatric age group. These are outlined in Table 3. To this group we can add a further case.

C. Case Presentation

A 4-year-old female presented with a three week history of back pain and deteriorating gait. Mild wasting of the buttock and right calf were noted on clinical examination. Plain X-rays of the spine showed a lumbosacral spina bifida and a myelogram revealed a mass extending from L_5–S_1. At surgery a

Table 3. *Paediatric Cases of Teratomatous Cysts of the Spinal Canal*

Author	Sex	Age	Level of lesion	Associated anomalies
Bucy and Buchanan 1935	m	32 mos	T_{11}–L_4	none
Hoefnagel *et al.* 1962	f	9 mos	$C_{5/6}$	none
Hyman *et al.* 1938	m	7 mos	C_1T_1	spina bifida
Ingraham, 1938	m	10 years	C_{5-7}	bifid T_{1-4} vertebral bodies
	m	2 years	T_9–L_1	spina bifida
	f	2 years	C_{2-5}	spina bifida
Masten 1940	f	5 years	C_4–T_1	none
Rewcastle 1964	m	15 years	L_1	none
	f	15 years	C_{5-6}	none
Larbrisseau 1980	m	4 years	L_1–L_3	none
H.S.C. 1985	f	4 years	L_4–S_1	spina bifida, diastematomyelia

large fluid containing cyst was found anterior to the cauda equina. In addition, there was a boney spicule enclosed by dura at L_5. Pathology of the cyst revealed a fibrocollagenous stroma supporting a singel layer of cuboidal cells consistent with a teratomatous cyst. A repeat myelogram 2 years later showed no evidence of recurrence.

5. Intraspinal Teratomas

A. Embryology

The various theories of histogenesis of the teratomatous cysts have also been used to explain the origin of sacrococcygeal teratomas. The origin of these tumors has been ascribed to primordial germ cells, embryonic nongerminal cells, extraembryonic cells and included maldeveloped twins. The various hypotheses are discussed in detail by Gonzales-Crussi[19].

B. Clinical Features

Teratomas are neoplasms composed of multiple tissues foreign to the part in which the tumor arises. Growth is progressive and incoordinated. Primary solid teratomas of the spinal cord would appear to be even less common than the rare teratomatous cysts. Garrison's recent review of primary intramedullary teratomas included only 9 cases of which 3 were children[18]. It is of interest that in two of these children the teratoma occupied the full extent of the spinal cord from foramen magnun to conus. There have been several case reports of teratomas found in association with myelomeningocoeles[40, 47] and in association with diastematomyelia and a lipomeningocoele[39, 55].

Sacrococcygeal teratomas are the most common teratoma in infancy and childhood occurring in 1 : 30 000–1 : 170 000 births[10]. Three percent of these teratomas can communicate with the subarachnoid space and directly involve the spinal cord[12]. We have had 2 cases of sacrococcygeal teratomas which directly and exentesively involved the spinal cord. One was malignant and the other benign.

C. Case Presentations

Case 1: A 7-year-old male presented with a 4 month history of back pain radiating into the right leg, a gradually progressive weakness of that limb and shortness of breath. He was found to be in cardiac failure and a loud bruit was auscultated over his sacrococcygeal area. Myelography showed an extradural mass extending below T_6 and a complete block of contrast media at T_{11}. Angiography revealed a vascular mass lesion. At surgery a sacrococcygeal tumor was identified within the pelvis and intraspinal extension noted. Histology determined this to be a malignant teratoma and the child died 3 months following surgery.

Case 2: A female infant was noted at birth to have a sacrococcygeal mass 10 cm in diameter and to be paraplegic with a sensory loss below T_4 (Fig. 5). Myelography demonstrated a block of contrast at T_3. Excision of the sacral mass showed tumor extending into the intraspinal space at a mid sacral level

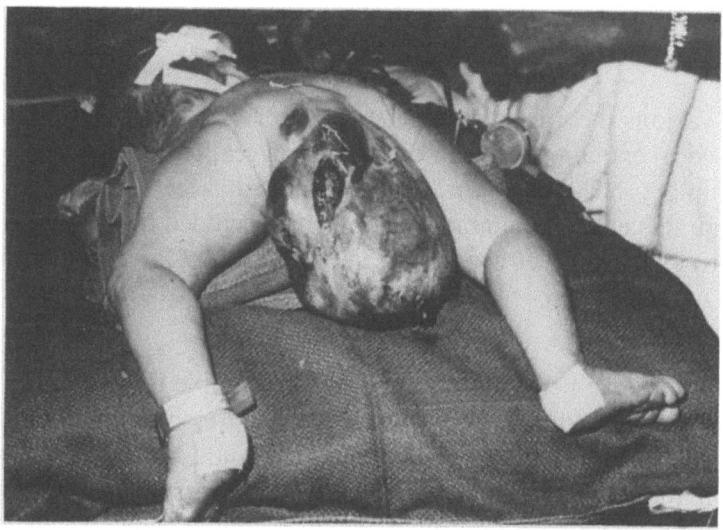

Fig. 5. Infant with a sacrococcygeal teratoma extending into a spinal cord to the level of T_2

and a subsequent T_4–L_3 laminectomy revealed intramedullary tumor extending to T_2. The extramedullary tumor was resected but an intramedullary removal was not attempted. In spite of residual tumor the child has done reasonably well. She is now 13 years old and able to ambulate with a walker. A re-exploration of cord in 1981 revealed intramedullary cysts but no residual tumor. A recent MRI scan shows a distended, enlarged cord suggesting, but not confirming, tumor presence.

6. Lipomas

A. Embryology

Early authors considered lipomas to be inclusion tumors, resulting from embryonic rests sequestrated within the cord. Chiari and von Recklinghausen reported fat cells appearing within pia arachnoid and theorized that lipomas may arise from these normally situated cells[1]. Ehni and Love postulated that in the process of vascularization of the cord, mesenchymal cells with the inherent potential for fat differentiation may invade the cord and give rise to fatty tumors[14].

More recently, it has been proposed that lipomas, in particular the lumbosacral lipomas associated with spina bifida, arise from the caudal cell mass. The caudal cell mass, as outlined by Alvord and by Lemire, extends in the primitive spinal cord from the posterior neuropore to the sacrococcygeal level[1, 37]. This mass regresses and differentiates to form the terminal ventricle, filum terminale and coccygeal medullary vestige, and contains pluripotential cells. Both of the above authors suggest that lipomas might arise from a clone of lipomatous cells that have escaped the normal control mechanisms during retrogressive differentiation. Walsh and Markersbery noted a variety of other tissue types in these lipomas which they feel could have arisen from other unsuppressed pluripotential cells during this same process[60]. This theory does not explain the occurrence of lipomas at sites other than the lumbosacral region, nor does it explain the occurrence of the occasional lipoma confined to the extradural space.

B. Clinical Features

Intraspinal lipomas unassociated with spina bifida are reported to comprise about 1% of all spinal cord tumors[14]. The more common lipomatous malformations associated with spinal dysraphism have been extensively reviewed in the paediatric literature, and the experience with these neuro-lipomatous lesions in this institution has been published[26, 36, 52].

A comprehensive review by Guiffre of nondysraphic intraspinal lipomas indicates that less than 20% of these lesions will presente for surgery in children less than 15 years of age, although in many adults the preoperative duration of symptoms extends into childhood[20]. The authors comment that a long clinical history is one of the signs suggesting an intradural spinal lipoma. In addition, spasticity and cutaneous sensory loss are almost invariable. These lesions, unlike the dysraphic lipomatous malformations have a predilection for the cervical and thoracic cord, and may, in some cases, involve the entire length of the cord. Intramedullary spinal lipomas can be accurately diagnosed by CT scanning. Total excision is difficult and may not be necessary for alleviation of symptomatology; so long as the bulk of tumor mass is resected. Both Guiffre and Woltman concur that the extent of removal does not influence long-term results[20, 64].

C. Case Presentations

We have had 2 cases in which an intramedullary lipoma involved the cervical cord, and in both cases the tumor extended into the brain stem.

194   H. J. Hoffman *et al.*:

Case 1: An 8 month old male infant was noted to be quadriparetic since 3 months of age. His cranial nerve examination was unremarkable and there was no clinical evidence of hydrocephalus. A CT examination revealed an intramedullary low density lesion extending the entire length of cervical cord and into the medulla to the level of the internal auditory meatus (Fig. 6). There was no contrast enhancement of the lesion.

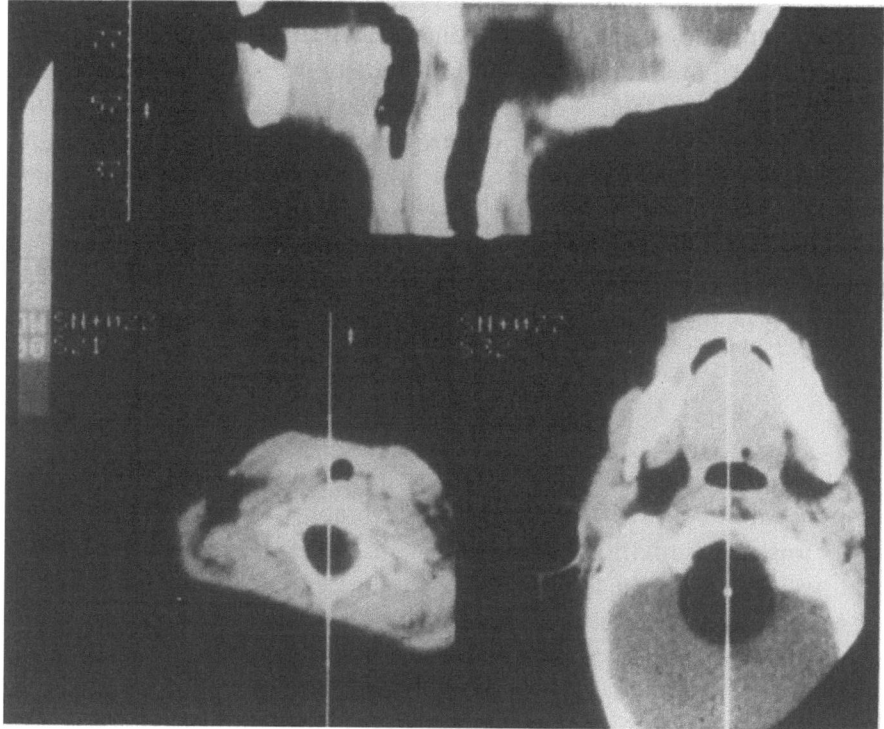

Fig. 6. Upper cervical and posterior fossa CT images of an intraspinal lipoma. Sagittal reconstruction demonstrates the extension of the tumor into the brain stem

A laminectomy revealed a continous intramedullary lipoma extending from T_2 to the medulla. The lesion was debulked with the ultrasonic aspirator. Post-operatively the child regained limb movements, and follow-up to date has shown improving function.

Case 2: A 10½-year-old boy presented with progressive spasticity of his lower extremities. He had a thoracolumbar scoliosis since infancy and a subcutaneous cervical lipoma had been noted since the child was 4 years of age. On examination he had a marked lumbar lordosis, a 5 × 5 cm subcutaneous midline lipoma and a large (60.5 cm) head. Tone and reflexes were increased in the lower limbs. A myelogram revealed an intramedullary tumor of the cervical cord with widening of spinal cord from the foramen magnum to C_4.

At surgery no connection was found between the subcutaneous mass and the spinal canal. An intramedullary lipoma was found extending from C_4 into the medulla and into the lumen of the IVth ventricle. The tumor was partially resected and pathology confirmed it to be a lipoma.

The patient has had a 15 year follow-up. A second decompression was performed a year after the first when the patient suddenly deteriorated. The patient was left with a spastic quadraparesis following the second procedure. He has however maintained bowel and bladder control, is able to drive a car and now teaches at a local university.

7. Spinal Arachnoid Cysts

Extradural spinal arachnoid cysts, alternatively termed intraspinal meningocoeles, represent a variant of spina bifida occulta as suggested by James and Lassman[32]. These cysts are enclosed by membranes containing both collagenous dural and arachnoidal elements. Narrow pedicles communicate the cysts with the subarachnoid space, and the cysts expand as pulsatile CSF is driven unidirectionally through this channel[13]. Wilkins collected 115 cases of extradural spinal cysts reported between 1898 and 1970, and noted that they occur primarily in younger patients[62]. Cloward and Bucy reported 10 cases of which 7 were less than 15 years of age[9].

Two-thirds of the adolescent patients with these cysts will present with Scheureman's kyphosis dorsalis and a number of authors have stressed that the diagnosis of an extradural arachnoid cyst should come to mind in any child with an unexplained kyphosis[9, 46, 63]. Neurological deficits may also be produced as the cysts gradually expand with CSF and compress the adjacent dural sac and cord[15, 24].

Scalloping of the posterior vertebral bodies is common on plain films and typically the myelogram shows an extradural filling defect at the level of the cyst. The cyst itself may eventually fill with contrast thereby confirming the diagnosis. CT metrizamide scans will distinguish the intrathecal sac from the extradural cyst (Fig. 7). Surgical excision of the sac is unnecessary so long as the communicating neck is obliterated. Surgery should effect a complete cure providing the associated cord compression has not been too long or too severe[62].

Intradural arachnoid cysts are occasionally found in patients with repaired myelomeningocoeles and most probably arise secondarily to postoperative arachnoiditis. We have had several patients develop symptoms from these cysts which resolved following cyst drainage. Rarely an intradural arachnoid cyst may be congenital however, as the following case illustrates.

Case 1: An 8-year-old boy with a lumber myelomeningocoele died of sequelae from his Arnold-Chairi malformation. At autopsy a non-cummunicating arachnoid cyst was found extending anterior to the cord from C_8 to L_1. The anteroposterior diameter of the cord was considerably flattened by the cyst, and the nerve roots were draped over its lateral extensions.

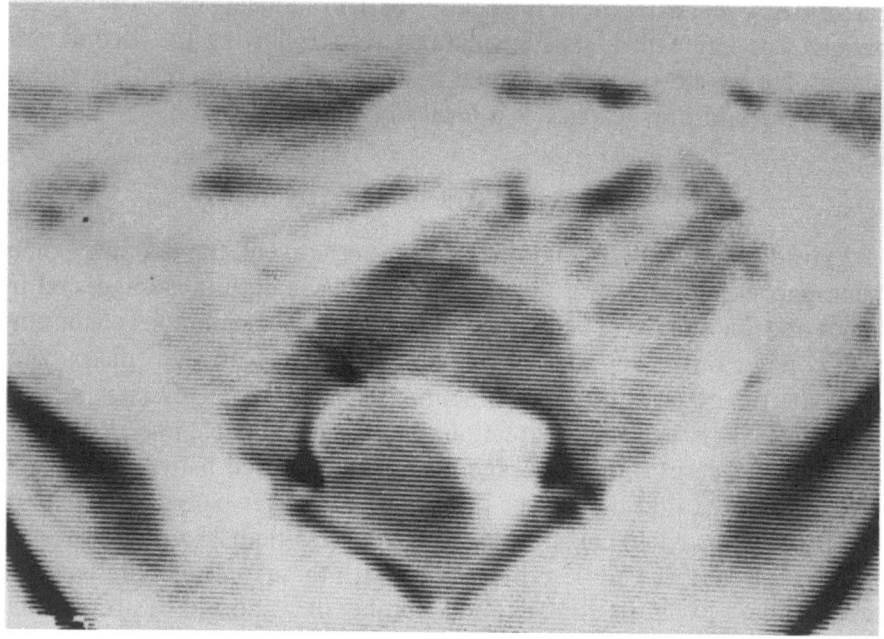

Fig. 7. CT scan of a metrizamide filled intraspinal meningocoele displacing and compressing the thecal sac

8. Congenital Malignant Tumors of the Spine

Primary CNS congenital malignancies are presumed to arise from embryonal cells in utero and include primitive neuroectodermal tumors, retinoblastomas, glioblastomas and neuroblastomas[51]. Although congenital intracranial neoplasms have been estimated to constitute approximately 1.5% of childhood brain tumors, primary congenital malignancies arising within the spinal cord have not been reported in the neonatal period[31, 54, 57].

Extradural compression of the cord by neuroblastoma is relatively common in infants. We have had 3 infants less than 5 months of age with spinal cord involvement by neuroblastoma and another 2 less than a year of age with similar pathology. Two of the infants had symptoms dating to birth; one child had hypotonic lower limbs and rectal prolapse at birth and the other a congenital monoplegia of her right leg. All five children underwent laminectomies and extradural excision of their tumors, followed by radiotherapy and in 4 instances, chemotherapy. All the children are alive and ambulatory four to sixteen years following their diagnosis, although 2 have neurogenic bladders.

The good outcome in these infants supports an aggressive surgical and radiotherapeutic approach inspite of longstanding lower limb weakness in several of the infants.

9. Summary

The paediatric spinal cord may host a wide variety of congenital tumors, the most common of which have been discussed. Although the majority of these lesions are benign their delayed diagnosis, serious infectious complications or associated congenital anomalies may expose the patient to life treatening morbidity or serious permanent disabilities. Prompt surgical management provides a cure in most instances, and the gratification of a neurologically intact child.

References

1. Alvord EC, Sham C-M, Sumi SM (1978) Central nervous system developmental abnormalities. In: Thompson RA, Green JR (eds). Pediatric neurology and neurosurgery. Spectrum Publications, Jamaica, New York, pp 97–166
2. Beardmore HE, Wiglesworth FW (1958) Vertebral anomalies and alimentary duplications: clinical and embryological aspects. Pediatr Clin North Am 457–474
3. Bentley JFR, Smith JR (1960) Developmental posterior enteric remnants and spinal malformations. The split notochord syndrome. Arch Dis Child 35: 76–86
4. Bostroem E (1897) Über die pialen Epidermoide, Dermoid und Lipoma und duralen Dermoide. Zentbl Allg Path Path Anat 8: 1–98
5. Boyd HR (1952) Iatrogenic intraspinal epidermoid. J Neurosurg 24: 105–107
6. Bremer JL (1952). Dorsal intestinal fistula: Accessory neurenteric canal; diastematomyelia. Arch Pathol 54: 132–138
7. Bryant H, Dayan AD (1967) Spinal inclusion dermoid cyst in a patient with a treated myelocystocoele. J Neurol Neurosurg Psychiatry 30: 182–184
8. Bucy PC, Buchanan DN (1935) Teratoma of the spinal cord. Surg Gynecol Obstet 60: 1137–1144
9. Cloward RB, Bucy PC (1937) Spinal extradural cyst and kyphosis dorsalis juvenilis. Am J Roentgenol 38: 681–706
10. DiLorenzo N, Griffre R, Fortuna A (1982) Primary spinal neoplasms in childhood: analysis of 1 234 published cases (including 56 personal cases) by pathology, sex, age and site. Differences from the situation in adults. Neurochirurgia 25: 153–164
11. Dodge HW, Keith HM, Campagna MJ (1956) Intraspinal tumors in infants and children. J Int Coll Surg 26: 199–215

12. Donnellan WA, Swenson O (1968) Benign and malignant sacrococcygeal teratomas. Surgery 64: 834–846
13. Duncan AW, Hoare RD (1978) Spinal arachnoid cysts in children. Radiology 126: 423–429
14. Ehni G, Love JG (1945) Intraspinal lipomas: report of cases, review of the literature and clinical and pathological study. Arch Neurol Psychiatry 53: 1–28
15. Elsberg CA, Dyke CG, Brewer ED (1934) The symptoms and diagnosis of extradural cysts. Bull Neurol Inst NY 3: 395–417
16. Fabinyi GCA, Adams JE (1979). High cervical spinal cord compression by an enterogenous cyst. J Neurosurg 51: 556–559
17. French BN (1982) The embryology of spinal dyraphism. Clin Neurosurg 30: 295–340
18. Garrison JE, Kasdon DL (1980) Intramedullary spinal teratoma: Case report and review of the literature. Neurosurgery 7: 509–512
19. Gonzalez-Crussi F (1982) Extra-gonadal teratomas, Fasc 18/2nd ser. Armed Forces Institute of Pathology, Washington
20. Guiffre R (1966) Intradural spinal lipomas. Review of the literature (99 cases) and report of an additional case. Acta Neurochir (Wien) 14: 69–85
21. Harriman DGF (1958) An intraspinal enterogenous cyst. J Pathol Bacteriol 75: 413–419
22. Hendrick EB (1982) Spinal cord tumors in children. In: Youmans JR, Neurological surgery, 2nd ed. W.B. Saunders, Co., Philadelphia
23. Hoefnagel D, Benirschke K, Duarte J (1962) Teratomatous cysts within the vertebral canal. Observations on the occurence of sex chromatin. J Neurol Neurosurg Psychiatry 25: 159–164
24. Hoffman GT (1960) Cervical arachnoidal cyst. J Neurosurg 17: 327–330
25. Hoffman HJ, Holness RO, Flett NR (1977) Long-term control of recurrent cyst of the conus medullaris. Case report. J Neurosurg 47: 953–954
26. Hoffman HJ, Taecholarn C, Hendrick EB, Humphreys RR (1985) Management of lipomyelomeningocoeles. Experience at the Hospital for Sick Children, Toronto. J Neurosurg 62: 1–8
27. Holmes GL, Trader S, Ignatiadis P (1978) Intraspinal enterogenous cysts. A case report and review of pediatric cases in the literature. Am J Dis Child 132: 906–908
28. Hosoi K (1931) Intradural teratoid tumors of the spinal cord. Arch Path 11: 875–883
29. Hyman I, Hamby WB, Sanes S (1938) Ependymal cyst of the cervicodorsal region of the spinal cord. Arch Neurol Psychiatry 40: 1005–1012
30. Ingraham FD (1938) Intraspinal tumors in infancy and childhood. Am J Surg 39: 342–376
31. Iooma R, Hayward RD, Grant DN (1984) Intracranial neoplasms during the first year of life: analysis of one hundred consecutive cases. Neurosurgery 14: 3–41
32. James CC, Lassman LP (1981) Spina bifida occulta. Orthopaedic, radiological and neurosurgical aspects. Academic Press, London
33. Klump TE (1971) Neurenteric cyst in the cervical spine canal of a 10-week-old boy. J Neurosurg 35: 472–476

34. Kubie LS, Fulton JF (1928) A clinical and pathological study of two teratomatous cysts of the spinal cord containing mucus and ciliated cells. Surg Gynecol Obstet 47: 297–311
35. Larbrisseau A, Renevy F, Brochu P, Decarie M, Mathieu J (1980) Recurrent chemical meningitis due to an intrapsinal cystic teratoma. Case report. J Neurosurg 52: 715–717
36. Lassman LP, James CC (1967) Lumbosacral lipomas: critical survey of 26 cases submitted to laminectomy. J Neurol Neurosurg Psychiatry 30: 174–181
37. Lemire RJ, Laeser JD, Leech RW et al. (1975) Normal and abnormal development of the human nervous system. Harper and Row, Hagerstown, Md, p 421
38. Masten MG (1940) Teratoma of spinal cord. Arch Pathol 30: 755–761
39. Mickle JP, McLennan JE (1975) Malignant teratoma arising within a lipomeningocele. Case report. J Neurosurg 43: 761–763
40. Mitang RN (1972) Teratoma occurring within a myelomeningocoele. Case report. J Neurosurg 37: 448–451
41. Oblu N (1967) Experimental investigation of the origin of intraspinal epidermoid cysts. Acta Neurol Scand 43: 79–86
42. Odake J, Yamaki T, Naruse S (1976) Neurenteric cyst with meningomyelocele: Case Report. J Neurosurg 45: 352–356
43. Padget DH (1970) Neuroschisis and human embryonic maldevelopment: New evidence on anencephaly, spina bifida and diverse mammalian defects. J Neuropathol Exp Neurol 29: 192–216
44. Padget DH (1968) Spina bifida and embryonic neuroschisis—a causal relationship. Definition of the postnatal conformations involving a bifid spine. Johns Hopkins Med J 123: 233–252
45. Palma L, DiLorenzo N (1976) Spinal endodermal cysts without associated vertebral or other congenital abnormalities. Report of four cases and review of the literature. Acta Neurochir (Wien) 33: 283–300
46. Raja IA, Hankinson J (1970) Congenital spinal arachnoid cysts. Report of two cases and review of the literature. J Neurol Neurosurg Psychiatry 33: 105–110
47. Reid SA, Mickle JP (1985) Myelomeningocoele occurring within a lumbosacral teratoma: case report. Neurosurgery 17: 338–340
48. Rewcastle NB, Francoeur J (1964) Teratomatous cysts of the spinal canal: with "sex chromatin" studies. Arch Neurol 11: 91–99
49. Rhaney K, Barclay GPT (1969) Entergenous cysts and congenital diverticula of the alimentary canal with abnormalities of the vertebral column and spinal cord. J Pathol Bacteriol 77: 457–471
50. Rosenbaum TJ, Soule EH, Onofrio BM (1978) Teratomatous cyst of the spinal canal. J Neurosurg 49: 292–297
51. Rubinstein LJ (1972). Presidential address: Cytogenesis and differentiation of primitive central neuroepithelial tumors. J Neuropathol Exp Neurol 31: 7–26
52. Swanson HS, Barnett JC (1962) Intradural limpomas in children. Pediatrics 29: 911–926
53. Takaku A, Kodama N, Ohara H, Hori S (1978) Brain tumor in newborn babies. Childs Brain 4: 365–375

54. Tomita T, McLone DG (1985) Brain tumors during the first twenty four months of life. Neurosurgery 17: 913–919
55. Ugarte N, Gonzalez-Crussi F, Satelo-Avila C (1980) Diastematomyelia associated with teratomas. Report of two cases. J Neurosurg 53: 720–725
56. Van Gilder JC, Schwartz HG (1967) Growth of dermoids from skin implants to the nervous system and surrounding spaces of the newborn rat. J Neurosurg 26: 14–20
57. Venes JL (1985) A proposal for management of congenital brain tumors. Concepts Pediat Neurosurg 6: 25–36
58. Wakai S, Arai T, Nagai M (1984) Congenital brain tumors. Surg Neurol 21: 597–609
59. Wakai S, Chiu C-W (1984) Rare combination of spinal lesions and spina bifida occulta: Case Report. Dev Med Child Neurol 26: 117–121
60. Walsh JW, Markesbery WR (1980) Histological features of congenital lipomas of the lower spinal canal. J Neurosurg 52: 564–569
61. Walker AE and Bucy PC (1934) Congenital dermal sinuses; a source of spinal meningeal infection and subdural abscesses. Brain 57: 401–421
62. Wilkins RH (1985) Intraspinal cysts. In: Wilkins RH, Rengachary SS (Eds), Neurosurgery, McGraw-Hill, New York
63. Wise BL, Foster JJ (1955) Congenital spinal extradural cyst. J Neurosurg 12: 421–427
64. Woltman HW, Kernohan JW, Adson AW, McCraig W (1951) Intramedullary tumors of the spinal cord and gliomas of intradural portion of filum terminale. Fate of patients who save these tumors. Arch Neurol Psychiatry 65: 378–393

Controversial Views of Editorial Board on the Intraoperative Management of Ruptured Saccular Aneurysms

Introduction

There are still a number of problems in fairly routine neurosurgery on which different views may be expressed. The editorial board of Advances and Technical Standards in Neurosurgery decided in December 1985 to publish the views of its members, all experienced surgeons, on a variety of subjects and the initial choice fell upon saccular aneurysms. This benign lesion can be successfully operated upon but there are many differing views about the best methods of approach.

This chapter summarizes the response to 15 questions raised by the board, each answered by a board member without the knowledge of the answers of his colleagues.

The manuscripts were collected by Professor Pertuiset of Paris, and have been edited by the Chief Editor.

1. What Is the Place of Lumbar CSF Drainage in the Operative Management of Aneurysms'?

Of the board members Brihaye and Guidetti regularly placed lumbar CSF drainage before the operation and opened the drain at the time of opening the dura. They found this an efficient and simple method to obtain a slack brain and reduce retraction. Guidetti further commented that the drainage can be discontinued, however, when the basal systems are available for aspiration.

Miller and Symon commented that they used lumbar CSF drainage only selectively, Miller for basilar aneurysms which he approached supratentorially. Symon operated in the lateral position, in upper basilar, posterior cerebral or posterior circulation aneurysms. Neither routinely used drainage for anterior circle aneurysms.

Loew was prepared to use CSF drainage in acute operations especially where there was a lot of subarachnoid blood but he found that the drain seldom function properly under these circumstances. In delayed or late operations he felt the use of drainage unnecessary since CSF was available from the basal systems.

Pertuiset, Nornes, Yaşargil and Pásztor did not use CSF drainage. Pásztor and Pertuiset were concerned that spinal drainage in the presence of some degree of brain swelling or raised intracranial pressure might provoke brain herniation. Pásztor commented that the sudden removal of CSF before dural opening might cause sudden decrease of ICP and increase of transmural pressure through the aneurysm thereby leading to premature rupture. He further commented that the presence of some CSF aided the dissection of the aneurysm particularly where there were adhesions round its neck, or in the dissection of middle cerebral aneurysms in the Sylvian fissure. Pertuiset felt that in the presence of an intraoperative haemorrhage a lumbar drain would inevitably become blocked whereas if there was no drainage the blood could be sucked directly out of the operative cavity. Yaşargil further noted that CSF could be obstructed within the ventricular system by haematoma and that in the presence of communicating hydrocephalus the CSF collected in the prepontine and intrapeduncular cisterns and was readily accessible to opening of Lilliequist's membrane which was a more effective method of reducing the intracranial contents than lumbar puncture. Pertuiset was further concerned about the possibility of the introduction of infection.

It appears, therefore, that there is no consensus about the routine use of CSF drainage but many people are prepared to undertake it without concern and others are prepared to undertake it in selected circumstances.

2. In the Presence of Considerable Dural Tension and no Evidence of Hydrocephalus on the CT, what Should Be Done? Should the Dura Be Opened?

Brihaye, Symon, Yaşargil and Pertuiset found this to be a very uncommon circumstance. Pertuiset commented that this had happened on 3 occasions all with patients in clinical grade 1; the explanation was obscure, and he had then closed the wound; he reoperated after 21 days. Brihaye had on a small number of occasions had to resect brain but felt that this constituted an error in judgement and could not be relied on to ease intracranial tension. Nornes, Pásztor and Yaşargil made the point that rerupture of the aneurysm during induction could of course, produce considerable tension and under these circumstances it might be impossible to proceed.

Guidetti, Miller, Symon and Pásztor noted that possible causes besides rerupture of the aneurysm were CSF accumulation with high CSF pressure but no hydrocephalus, brain edema or space occupying hematoma. Pásztor was prepared to place an in-dwelling catheter in the frontal horn and then proceed with gentle brain retraction over protective rubber strips to reach the chiasmal cistern or, passing between the optic tract and the internal carotid artery, the interpenduncular cistern through Lilliquist's membrane.

Only Loew and Pertuiset had ever found it necessary to abondon surgery under these circumstances.

3. If Brain Swelling Is Present Should Brain Resection Be Performed?

Loew and Yaşargil never use brain resection. Pásztor and Pertuiset commented that where brain compliance was low and there was considerable brain swelling, resection of the temporal pole together with uncus and temporo-medial structures might tide the patient over a dangerous period of raised intracranial pressure. In the opinion of both these board members this was an uncommon situation. It might arise in the course of an operation quite quickly and some resection might be necessary to complete the operation or to remove hematoma.

Pertuiset and Symon have resected portions of the anterior temporal lobe routinely in the approach to upper basilar aneurysms, Symon and Guidetti used small gyrus rectus resections routinely in the approach to anterior communicating aneurysms and Symon occasionally used wedge resection of the medial frontal lobe in handling giant aneurysms of the anterior communicating complex from a medial parasagittal approach.

Miller used brain resection either of the frontal or temporal lobe only as a last resort in handling some sort of operative disaster, and board members were unanimous in indicating that resection of major portions of the brain in routine aneurysm surgery was uncommon, unnecessary and quite often unhelpful.

4. In the Presence of an Aneurysm Hematoma Should It Be Removed Before Tackling the Aneurysm Neck?

Miller and Symon commented that they usually approach the aneurysm neck first before extensively evacuating the hematoma unless the hematoma was space occupying and approached on the way to the aneurysm when it would be cautiously evacuated leaving some haematoma round the fundus. This was also Brihaye's view and Guidetti commented that if the hematoma was huge, he would begin by partly evacuating it, if it were small, he would clip the neck first. Loew made the differentiation between aneurysms which were likely to bleed, that is operated on recently after hemorrhage when disturbance of the hematoma should be minimal, only enough to approach the aneurysm neck, and aneurysms which were soundly hemostased when evacuation of the hematoma could precede the clipping of the aneurysm in order to get better access and avoid pressure on the brain. This was also Nornes view and the view of Pásztor; only if the clot was very large would it be tackled first rather than after clipping the neck.

5. Is Operating on Aneurysms at the Present Time Justifiable Without a Microscope? What About Loupes?

Here there was unanimity of view; no members of the board felt that a loupe was a reasonable alternative to the operating microscope and felt that if operations on aneurysms were to be considered, then a microscope was essential. Miller and Loew commented that in days gone by, loupes and a headlight could be better than nothing, but Yaşargil pointed out the advantages of the small intrapupillary distance of the microscope, the sharp stereoscopic view in a deep field, (none of which were available with the loupe) and several members of the board had no experience with the use of operating loupes at all.

6. Is Self Retaining Retraction Absolutely Necessary?

This produced interesting answers; Brihaye, Loew and Pásztor commented that although a self-retaining retractor was not absolutely necessary it was extremely helpful, Pásztor making the point that if the brain were slack and the assistant experienced, hand retraction could be used. Nornes used self-retaining retraction for most aneurysms and also used a hand held retractor in cases of posterior inferior cerebellar artery aneurysms commenting that it was mandatory that the assistant had a clear view through the operating microscope. Guidetti, Miller and Symon regarded the use of a self-retaining retractor as absolutely mandatory in association with microscope surgery, Guidetti commenting that this avoided the obstruction of the field by the presence of an assistant's hand. Pertuiset regarded the use of a self-retaining retractor as essential and thought it should be fixed on the operating table whereas the others generally fixed the self-retaining retractor to the skull.

Only Yaşargil regarded self-retaining retraction as not necessary. He commented that if the brain was fully relaxed, most effectively through preliminary opening of the basal cisterns, retraction was unnecessary and at any rate should only be used for as short a time as possible to release CSF from the cisterns. He rarely used retraction in middle cerebral aneurysms and never 2 retractors. He thought that he could operate satisfactorily with only retraction of the pressure of a sucker tip over patties.

7. Should Mean Arterial Pressure Be Recorded and Should Measurements of Pulmonary Artery Pressure Be Made?

All the board members routinely recorded systemic arterial pressure and regarded mean arterial pressures as the most satisfactory on which to base clinical judgements. Pertuiset and Yaşargil in addition monitored pulmonary pressure using a Swann Ganz catheter but pulmonary artery pressure was not recorded by any of the other board members except, in Guidetti's case, patients with cardiac disease or where very deep hypotension was necessary. Loew commented that pulmonary artery pressure was required only in patients with major blood loss. Pásztor and Miller regarded central venous pressure as helpful and did not use pulmonary artery pressure.

8. What Is the Place of Arterial Hypotension and Temporary Clipping?

Controlled hypotension was used routinely by Guidetti, Pertuiset, Loew, Symon and Pásztor. All made the point that the degree of hypotension had to be determined by the age and general condition of the patient, Symon pointing out that the autoregulatory curve was shifted upwards in patients with hypotension and that the routine level of 60 mm Mercury mean, also used by Pertuiset, or 70–80 mean as used by Pasztor might be too low when the patient had hypertension or renal or cardiac insufficiency. Guidetti and Pertuiset were prepared to use mean arterial pressure down to 30–40 mm Mercury with nitroprusside in case of surgical necessity. Miller did not routinely use arterial hypotension but reserved its use only for dissection of a difficult sac or where there had been intraoperative rupture controllable only with a pledget over the hole in the aneurysm. Yaşargil would use hypotension only in cases of aneurysm of the basilar bifurcation and Nornes had abandoned routine systemic hypotension some years ago. He aimed to keep a stable blood pressure at or slightly below the preoperative blood pressure.

All the board members were prepared to use temporary clips, Guidetti and Symon commented on their utility in the dissection of giant aneurysms and Yaşargil emphasized that utility where the aneurysm was thin-walled and the parent vessels very adherent to the sac. Where premature rupture occurred, Miller, Loew and Yaşargil commented in the utility of the use of temporary clips and Symon pointed out that if the aneurysm were bleeding the distal as well as the proximal circulation should be excluded by clips to prevent profound hypotension in the collateral circulation. Guidetti routinely used hypertension while the temporary clips were applied and Brihaye found their use particularly helpful in anterior communicating and middle cerebral artery aneurysms. Nornes commented that he gave 7,500 IU heparin intravenously before occluding the feeding vessel of an aneurysm if the clip was expected to remain in place for more than 2 minutes and both he and Symon routinely used Scoville clips as temporary clips. Pásztor in general cautioned against the use of temporary clips because of the possible threat of damage to the intima and indicated that where it was necessary to use them it was important to avoid the possibility of occlusion of the parent vessel by the definitive clip on the aneurysm neck because the parent vessel had collapsed under the influence of a proximal clip. He, therefore, advised that if proximal clips had been used, the definitive clip on the neck be released for a moment to allow the parent vessel to fill up before the neck clip was reapplied. Pertuised used temporary clips only where profound hypotension was contraindicated.

9. Should Bipolar Coagulation Be Used to Reduce the Size of the Sac or to Reduce the Caliber of Its Neck?

Brihaye and Nornes never used bipolar coagulation on aneurysms. Loew used bipolar coagulation if the neck was large and difficult to close properly with a clip when coagulation might reduce its caliber. He did not use coagulation on the sac. Miller and Symon used coagulation both to reduce the size of the sac and to reduce the caliber of the neck when necessary; Miller commented that bipolar coagulation was in addition useful to occlude multiple tiny aneurysmal dilatations along the middle cerebral artery. Pásztor used bipolar coagulation only when temporary clips had been placed on the neck and required to be changed to a better definitive placement. In these cirumstances the fundus could be coagulated to aid the safety of the transfer of clips. He did not regard bipolar coagulation as a substitute for clipping and never coagulated areas which would not be clipped for the neck before clipping. Pertuiset used bipolar coagulation both on the neck and on the sac but never on necks less than a week after hemorrhage because of fear of perforation of soft tissues.

Guidetti indicated that in using bipolar coagulation the forceps tip had to be completely across the aneurysm neck and as long as the tips of the forceps were clean and unpitted, the current was low and the surgeon proceeded gradually, opening and closing the tips to prevent adhesion, he had never seen harm come from the procedure.

Yaşargil routinely used bipolar coagulation to reduce bulging parts of the aneurysm sac. He commented that he originally had used coagulation to reduce the caliber of the neck or to produce a neck in cases in which none was immediately discernible but had more lately changed the technique so that the arachnoid fibers at the base of the aneurysm were gently freed by blunt dissection and then a preliminary pilot clip applied to the neck. Thus protected he dissected the sac from adjacent vessels, identified the perforators, coagulated the fundus of the aneurysm down towards the pilot clip and then removed the pilot clip while the rest of the aneurysm sac was further coagulated down to its base and having reached normal anatomical layers applied a final clip. He had, occasionally, following application of the pilot clip resected a large fundus and coagulated it down to its origin. On other occasions he had used 2 pilot clips to a long neck, that nearer the parent vessel being then removed with coagulation of the base until application of a single final clip was possible when the second pilot clip could be removed. He commented that almost every aneurysm directed superiorly or posteriorly from a bifurcation might have an inferior bulging aspect which could remain hidden and difficult to deal with. The philosophy was as he said "first to get the wild horse with a lasso around its neck and then to tame it, but one must ensure that it is fully tamed". He commented that such techniques had limitations when the aneurysm was large and sclerotic.

10. What Is the Preferred Method of Occlusion of the Neck, Which Clip? Are Ligatures ever Necessary?

Brihaye and Yaşargil always used Aesculap clips, never ligatures. Loew usually used Yaşargil clips, more rarely Sugita clips. Miller used the McFadden variangle clip or Sugita and almost never ligature. In Loew's view ligature of the neck of a large aneurysm was a safe method but relatively seldom necessary now. Guidetti and Pásztor used the Aesculap, Sugita or Drake clips, Guidetti commenting that the clip must lie flush with the artery to avoid the development of further swelling from the aneurysm neck. Even if this required several repositions it was mandatory to do this. Occassionally he would reinforce a residual unclippable base with thin cotton to promote scar.

Pásztor commented that he had available only a few types of clip whereas Nornes had a complete range and would use any that appeared convenient. The Yaşargil, Sugita and Drake clips had proved most useful to Pásztor particularly where a vessel had to be left out of the blade and the Drake fenestrated clip was the only solution. He felt that ligature was never satisfactory. Symon used Scoville clips almost routinely having an instrument curator who was prepared to produce bent Scoville clips in various shapes. He also used Aesculap and Sugita clips and very rarely in the case of giant aneurysms ligated the neck.

The neck of giant aneurysms could occasionally be prepared for clipping by gentle crushing with Mosquito forceps as pointed out by Guidetti and this technique had also been used by Symon.

Pertuiset commented on the utility of new Caspar applying forceps, the 15° angled jaw and a complete 360° range.

11. Is Coating a Safe Procedure

The general view of the board was that coating by itself was an unsatisfactory procedure. Pertuiset indicated that he never used it, Brihaye had used it in a number of middle cerebral aneurysms but had had rebleeding following it. Guidetti regarded coating as possible only in small aneurysms, and used it routinely where some tiny aneurysms became apparent during surgery as a thin transparent bubble on an artery. These small lesions had no neck and would be dealt with by bipolar coagulation followed by coating.

12. What Should Be Done when an Intraoperative Rupture Occurs?

Guidetti, Nornes, Pásztor and Symon all indicated that the management depended on the stage of the operation of which rupture occurred. The worst possible case was where intraoperative rupture occurred before the aneurysm had been approached. There was no time to elevate even a bulging lobe and Pásztor might make a quick incision and a small basal resection controlling bleeding using a large sucker and attempting to clip the neck promptly. Symon would call for a rapid induction of lower hypotension and the application of a proximal arterial clip for example on the terminal carotid artery to attempt to slow flow through the aneurysm sufficiently to obtain control. He would then place a distal clip to prevent efflux from the collateral circulation, and dissect the neck under the protection of the two clips. Guidetti also found that temporary occlusion of feeding arteries was helpful in such early ruptures but preferred such occlusion for the briefest possible time. Nornes found the utility of the use of two suckers in such premature ruptures dissecting with one and controlling hemorrhage with the other. Pertuiset took a similar view, where rupture occurred before the neck of the sac was accessible, profound hypotension, the use of temporary clips or partial clipping of the sac enabling the operation to proceed. Yaşargil, Brihaye and Loew routinely used suction on the aneurysm over cottonoids with the use of temporary clips only when the bleeding could not be arrested by this method.

Later in the course of the operation Symon, Pertuiset, Pásztor and Guidetti all found that control of bleeding by gentle pressure over a patty with the loss of aneurysmal tension might in fact make dissection of the neck much more rapid and easy without the necessity for a temporary clip to the aneurysm or parent vessel.

All emphasized that rupture at the stage of aneurysm preparation was a far less serious event unless rupture occurred at the base of the aneurysm in which case temporary vascular occlusion would certainly be necessary. Yaşargil and Nornes commented that the application of a clip across the sac, might control a fundus rupture and Yaşargil and Pertuiset would then consider the use of bipolar coagulation on the neck to facilitate the application of a permanent clip.

13. What Should Be Done when there is Obvious Spasm of the Internal Carotid Artery at Operation?

Brihaye, Symon, Yaşargil and Loew were prepared to use papaverine on pledgets or in solution on the internal carotid artery. Loew in addition had used nimodipine and Miller, naftidrofuryl. Miller and Pertuiset commented on the use of copious irrigaton of warm saline, Pertuiset expressing little confidence in the use of papaverine, and commented that where spasm was present he did not use profound hypotension. Yaşargil favored careful dissection of the small sympathetic fibers on the outer wall of the vessel.

Nornes commented on his use of the pulsed Doppler machine for intraoperative measurement of blood flow velocity reflecting the degree of lumen reduction and had also used papaverine and stripping of the adventitia. He emphasized the necessity of avoiding marked systemic hypotension under these circumstances and the utility of increasing blood volume and careful checking of electrolytes.

Guidetti had found no agent capable of reversing severe vasospasm and both he and Pásztor indicated that the timing of surgery was of extreme importance in relation to the avoidance of vasospasm during operation. Guidetti felt that very early operation might avoid such problems while Pásztor would avoid operating when there was evident vasospasm on a preoperative angiogram.

14. What Is Your View of Cardiac Arrest and Hypothermia?

Brihaye, Pertuiset and Yaşargil had never used cardiac arrest or hypothermia. Guidetti, Symon and Loew had used hypothermia with surface cooling in the past, but had abandoned it some 20 years ago. Pásztor felt that extensive hypothermia, which he had also abandoned many years ago, and cardiac arrest were techniques which in themselves carried appreciable danger. Only Nornes in the editorial board had used cardiac arrest with cooling by exchange apparatus in association with a heart/lung machine and Symon commented that in rare circumstances this type of cooling as advocated by Silverberg in the United States of America and used by Rice-Edwards in England, might present advantages in extremely difficult aneurysms.

The consensus was that these techniques had little place in routine aneurysm surgery but might be helpful in very rare conditions.

15. What Should Be Done when a Junior Surgeon Calls for Help During an Operation

Pásztor and Miller stressed the value of graduated training before juniors operate on aneurysms by themselves. Miller commented that even when his assistants were regarded as fully competent to do the entire operation he almost always remained in theatre clothes within the operating theatre suite and asked to be called when the aneurysm was ready to be clipped or when a difficult dissection with possible rupture was under way. He commented that he had only once had to be called in to take over a disaster where the aneurysm had ruptured and bleeding could not be stopped. Pásztor's view was that when help was called for he would go at once since this reflected a situation serious enough for his colleague to feel stressed and to require support. Pertuiset and Symon would invariably take over the operation and finish it themselves, Pertuiset thereafter congratulating his junior for his judgement in self-assessment as being unable to cope with a difficult situation. A similar attitude was taken by Loew who would immediately go to theatre, listen to the problem and either give advice or take over. Nornes would send the message "Keep your temper, you can certainly manage" and at once start for the theatre.

In Yaşargil's view a junior neurosurgeon should not be allowed to perform such surgery till sufficiently dexterous and theoretically knowledgeable to carry out the procedure by himself. He stressed the value of a comprehensive study of neuroanatomy, a thorough grounding in possible technical maneuvers necessary to extract himself from difficulties and a

Table 1. *Answer to Questions*

		Yes	No	It depends of the situation
1	Lumbar drainage	2	5	2
2	Opening of a tense dura	5	3	1
3	Brain resection	2	2	5
4	Evacuation of hematoma first	7	2	—
5	Microscope	9	—	—
6	Self-retaining retractor	7	2	—
7	Recording of the mean arterial pressure	9	—	—
8	Arterial hypotension	7	2	—
	Temporary clips	8	1	—
9	Bipolar coagulation for reducing the sack	7	2	—
10	Clip for occlusion of the neck	9	—	—
11	Coating	1	7	1
14	Cardiac arrest and hypothermia	—	9	—

Unanimity 9	4	9
Positive agreement 7	5	
Controversis 5	3	

thorough laboratory training in vascular surgical technique. Guidetti felt that as an older surgeon it was always his duty to go to help and Brihaye had never experienced the need to actually take over from a younger surgeon.

Conclusion

A summary of the views of the editorial board to 12 questions of the 15 is given in Table 1. Questions 12, 13, and 15, were impossible to summarize. In general, there was fair unanimity between this group of European surgeons on the handling of the majority of crises affecting aneurysm operations. It is hoped that the discussion of these issues may prove helpful and from time to time, Advances and Technical Standards will consider other controversies in neurosurgery. Younger surgeons are invited to communicate with the Editorial Board to raise particularly interesting questions which they would like to see discussed.

Author Index

Abe H 77, 118
Achor LJ 59
Ackerman R 31
Ackmann JJ 43
Adams JE 198
Adeniyi-Jones R 91
Adler G 49, 59
Adson AW 193
Akikawa K 129
Alder VG 21
Aldor AM 115
Alexander AN 98
Alexander L 89
Alexander N 86
Allison T 44
Alvord EC 192, 193
Ambrosi B 77, 83
Amoss M 74, 75
Amr S 80, 81
Anderson DC 86, 89
Anderson MS 84
Anderson PE 26
Arai T 176
Arden G 55
Arimura A 74, 84, 103, 113, 114
Astrup J 26, 27
Avezaat C 43
Avila A 60

Baba Y 74, 84
Baetens D 113, 115
Baird A 101, 102, 103, 105
Barclay GPT 186, 189
Barnett JC 193
Bassiri R 75
Basso A 171
Bauer W 117
Bauman B 127

Beardmore HE 186
Beardsworth DE 88, 92
Beck-Peccoz P 77, 80, 81, 83
Becker DP 43, 44, 47
Becker H 86, 89
Belanger A 93
Bell BA 27
Benirschke K 189, 190
Ben-Jonathan N 74, 118
Benett GW 75, 77, 78, 79, 83
Benoit FR 103
Benoit R 103
Benraad THJ 100
Bentley JFR 186
Ben-Zeev Z 127
Berci G 6
Berelowitz M 106, 113
Berg D 91, 92
Berger F 111
Berger G 111
Berger L 51
Bergquist C 88, 92
Bernutz C 80
Bertelink AKM 100
Besser GM 86, 89, 103, 105, 112
Beyer J 117
Bhatt RS 74, 102
Bickford RG 51
Bier D 117
Bigos ST 79
Bilezikjian L 74, 94, 95, 96, 97
Billinger TW 51
Bint Akhtar F 88
Bitensky L 80
Blackard WG 112
Blackwell R 74, 75
Blecher PE 85
Blizzard RM 108, 109

Bloch B   103, 111
Bloom F   74, 94, 95, 96, 97, 103
Bloom SR   105, 113, 117
Bock L   105
Bode HH   88, 92
Böhlen P   74, 101, 102, 103, 105
Bollinger-Gruber J   77, 118
Bond M   45
Bonnal A   171
Bonnal J   171
Borges JLC   105, 108, 109
Borowski GD   78
Boston JR   63
Bostroem E   177
Botalla L   77, 90, 111
Bourguignon JP   86, 89
Boyd AE   112
Boyd HR   178
Boysen G   26
Braakman R   43
Branston NM   26, 27, 35
Braunstein GD   77, 115
Brazeau P   74, 101, 102, 103, 105, 111, 112
Bremer JL   185
Brenner MJ   113
Brewer CC   59
Brewer ED   195
Brierley J   27
Briger U   117
Brihaye J   201 ff
Brochu P   190
Brock M   17
Brooks EB   59
Brown AH   21
Brown GM   92
Brown MR   74, 94, 95, 96, 97
Brownstein MJ   75
Bruce DA   59
Bruce DL   55
Bryant H   178
Buchanan DN   189, 190
Bucy PC   177, 189, 190, 195
Bull JW   31
Bunnell WP   64
Burger HG   75, 79, 93
Burgus R   74, 75, 112

Burrin M   117
Butcher M   74, 75

Calvet J   51
Campagna MJ   176
Cant BR   28, 30, 43, 44
Carey RM   97, 101
Carlin J   43, 59
Carpenter PC   111
Carrizo A   171
Carter L   64
Casler JA   51
Castermans A   171
Cathala HP   51
Caton W   43
Catt KJ   85
Cauter E van   98
Chadal P   106
Chapman AJ   79
Charest NJ   112
Charlton BG   96
Chayvialle JA   111
Cheron G   28, 30
Chiappa KH   59
Chiodini PG   77, 90, 111
Chitwood J   109, 112
Chiu C-W   177
Ch'ng LJC   117
Christensen S   43
Chrousos GR   97, 100
Chud L   85
Ciganek L   54, 55
Claustrat B   111
Clayes L   156
Cloward RB   55, 195
Collier KJ   74, 102
Collins SM   115
Collins WP   86, 89
Colturi TJ   113, 114
Comaru-Schally AM   127
Comoy C   4
Condon EM   83
Cone J   31, 35
Conlon JM   115
Cook NA   160
Costa e Silva IE   54, 57
Coté J   94, 96, 97

Cowan JC 30
Coy DH 96
Coy EH 102
Crawford JD 88, 92
Crigler JF 88, 92
Crockard HA 26
Cronin MJ 101, 105, 112
Crowell RM 27
Crowley WF 86, 92
Culler MD 103, 114
Curcic M 60
Cushing H 137
Cutler GB 97, 100

Daughaday WH 77, 82, 83
Davis GR 115
Dayan AD 178
Dayle E 63
DeBold CR 98
Decarie M 190
DeCherney GS 98
Dee PC 74
DeJong F 97
Delmas A 171
Del Pozo E 77, 83, 90, 112, 117, 118
Demoulin A 86, 89
Demura H 95, 96, 97, 103, 106, 109, 114
Derome J 171
Derome P 171
Desiderio D 74
Desmedt JE 28, 30
Dettmar PW 84
Deuben RR 101
Dhariwal APS 112
DiGeorge AM 79
DiLorenzo N 176, 189, 191
Dimaline R 118
Djndjian R 156
Docherty TB 55, 57
Dodge HW 176
Doelle GC 127
Doepfner W 117
Dolva LO 79
Dongen KJ van 117
Donnellan WA 191
Dörr HG 97, 98, 99, 100

Dosoretz DE 172
Downs T 112
Drake CR 97, 101
Draznin M 101, 112
Duarte J 189, 190
Dubois MP 113
Du Bouley GH 31
Dudley CA 85
Dufau ML 86
Duick DS 79
Duncan AW 195
Dunn TF 74
Dupont A 93
Dyke CG 195

Edgar MA 64
Edison 4
Edwards CRV 112
Ehni G 192, 193
Ehrenberger K 17
Ehrensing RH 84
Eichling JO 31
Eisenhardt L 137
Elias KA 118
Elsberg CA 195
Emson PC 115
Enas GG 43
Engel WK 84
Engler D 77, 118
Erhardt F 79
Ernould CH 86, 89
Esch F 74, 101, 102, 103, 105
Esch W 102
Evans LEJ 84
Evans WS 105, 106, 108, 109
Ezrin C 77, 115

Fabinyi GCA 198
Faden AI 84
Faglia C 80
Faglia G 77, 80, 81, 83
Fahlbusch R 79, 80, 83, 90, 112, 117
Farah S 156
Fawcett CP 85
Fay T 4
Fehm HL 119
Feinsod M 51, 55, 57

Feldman M   113, 114
Fellows R   74, 75
Ferguson GG   31
Ferin M   88
Ferrari C   77, 80, 83
Ferrie IN   96
Fischer C   172
Fischer G   172
Fitch W   31
Flett NR   184
Florsheim WH   114
Flückiger E   77, 83, 90, 117
Foley K   106
Foncin JF   173
Forest MG   86, 88
Forsham PH   117
Fortuna A   176, 191
Foster JJ   195
Fourestier M   4
Fournier A   4
Fox JE   28, 31
Franchimont P   86, 89
Francoeur J   189, 190
Freischem CW   88
French BN   177
Frohman LA   105, 106, 112
Fry J   43, 59
Fulton JF   188
Furlanetto R   106, 108, 109
Furutani Y   74, 94

Gado MH   31
Gage LP   74, 102
Galbraith S   43
Ganong WF   86
Garofano CD   78
Garrison JE   191
Gelato MC   108, 109, 112
Gennarelli TA   43, 59
Gentili F   39, 43
Gerich JE   112, 113, 115, 117
Gershengorn MC   79
Giguère V   94, 96, 97
Gildenberg PL   43
Gilder JC van   178
Gillespie WA   21

Gillies GE   97
Girod C   111
Girolami U de   27
Gold PW   79, 97, 100
Goldie WD   59
Gonzalez-Crussi F   191
Goodman RH   74
Goodman RS   127
Goodwin FK   79
Graham K   118
Grant DN   196
Grant FC   4
Grant H   43
Green RF   113
Greenberg RP   43, 44, 47
Griffith EC   75, 77, 78, 79, 83
Griffith HB   10
Griffre R   176, 191
Grimelius L   113
Groux R   4
Grubb RL Jr   31
Grumbach   87
Grundy BL   43, 63
Gubler U   74, 102
Guidetti B   201 ff
Guiffre R   193
Guillemin R   73, 74, 75, 101, 102, 103, 105, 112
Guiot G   4
Guyot JF   172

Haase J   43
Habener JF   74
Habermann J   79
Hackens WHL   97
Hagen B   97, 98, 99, 100
Hall R   84, 86, 89, 114
Haller R   117
Halliday AM   52
Halliday E   52
Hamby WB   190
Handa J   35, 43
Hankinson J   195
Hansmann M   90, 91
Harding GFA   54
Harper AM   31
Harriman DGF   186

Harringtom T 43
Harris GW 73
Harris RJ 27
Harsoulis P 86, 89
Hartl R 105, 108
Hartwimmer J 96, 97, 98, 99, 100
Hattori S 64
Hayakawa T 26
Hayward RD 196
Heath M 114
Heiden J 43
Hein O 43
Heinze E 119
Heiss WD 26
Hellman P 112
Hendrick EB 176, 193
Henry KR 60
Hermus ARMM 100
Heros RC 43
Hess RM 31
Hetzel WD 119
Hinkle PM 77
Hirose T 51
Hirsch J 51
Hizuka N 109
Hoare RD 195
Hockley AD 79
Höfer R 79
Hoefnagel D 189, 190
Hoffman BJ 74, 102
Hoffman GT 184, 195
Hoffman HJ 193
Hoffman R 51
Hofmann AR 86, 92
Holaday JW 84
Holl R 119
Holmes GL 188
Holness RO 184
Homma S 64
Hopkins HH 5
Hori S 176
Horide R 5
Horn K 79, 80
Horvath E 77, 101, 112, 115
Hoso E 103
Hosoi K 189
Hostetter G 85

Hotta M 106, 114
Hoyt WF 51
Huguenin R 117
Hume AL 28, 30, 43, 44
Humphrey PRD 31
Humphreys RR 193
Hunt WE 31
Hunter P 84
Hvid-Hansen O 43
Hyatt MS 43
Hyman I 190

Ignatiadis P 188
Imahi T 97, 106
Imai K 64
Imaki I 103
Imaki T 114
Imura H 118
Ingraham FD 190
Inoue S 64
Inoue SI 64
Inoue T 118
Iooma R 196
Ishikawa T 5
Island DP 98
Ituarte E 112
Ivell R 74
Iversen LL 115

Jackson IMD 75, 77, 84, 89, 119
Jackson RV 98
Jacobs HS 91
Jacobs JW 74
Jacobs LS 77, 79, 82, 83
Jacobs TP 84
Jacobson JH 51
Jänicke F 91, 92
Jaksche H 15
Jakubowsky J 57
James CC 193, 195
James IM 31
Jane JA 43
Jennett B 31, 43, 45
Jewkes D 39
Jibiki K 97, 106

Joffe SN 106
Johanson A 101, 112
Johanson ML 86
Johnston M 84
Jones BR 55
Jones SJ 64
Jones TA 60
Joplin GF 117

Kabayama Y 118
Kaiser DL 105, 106, 108, 109
Kaliebe T 96
Kálmánchey R 60
Karam J 117
Kasdon DL 191
Kastin AJ 84
Kato Y 118
Kawai S 64
Keith HM 176
Kelsey JC 21
Kempe GL 166
Kennedy I 43, 59
Kerdelhue B 85
Keret R 127
Kernohan JW 193
Kewenig M 80
Kidooka M 35, 43
Kirsch WN 51, 55, 57
Kirschner F 127
Kivelitz R 49, 59
Kiyosawa Y 97, 103, 106
Kloppenborg PWC 100
Klump TE 189
Knobil E 85
Knuth UA 88
Kodama N 176
König A 119, 120
Kohner EM 106
Kolbe H 79
Kont LA 6
Koob G 74, 94, 95, 96, 97
Kotani H 64
Kourides IA 79
Kovacs K 77, 97, 101, 112, 115
Kozbur X 114
Kozlowski GB 85
Kraenzlin ME 117

Krause U 117
Krejs GJ 115
Krieger DT 83, 97, 98, 112
Kriss A 51, 52
Krulich L 112, 114
Kubie LS 188
Kubo M 129
Kurze T 43
Kushner DV 88, 92

Labrie F 93, 94, 96, 97
Ladds A 35
Laeser JD 193
Lamberts SWJ 97, 117
Lance VA 102, 109
Land AM 43
Lange E 49, 59
Langfitt TW 43, 59
Larbrisseau A 190
Laron Z 127
Larson SJ 43
Lassen NA 26, 27
Lassman LP 193, 195
Leake A 96
Leech RW 193
Lefèvre G 96
Legros JJ 86, 89
Lehman RAW 51, 55, 57
Leisner B 79
Leitner JW 102, 114
Lemire RJ 193
LeRoith D 113
Lesbros F 111
Lesoin F 171
L'Espinasse VL 3, 4
Leong DA 112
Leuschner U 116
Levine RA 61, 62, 63
Lewis M 114
Lewy RA 78
Leyendecker G 90, 91, 92
Lightman S 118
Lina A 43, 63
Linde R 127
Lindsay KW 43, 59
Ling N 74, 75, 97, 101, 102, 103, 105, 106, 109, 112, 114

Linton EA 96, 97
Liuzzi A 77, 90, 111
Loew F 15, 49, 59, 201 ff
Lomedico PT 74, 102
London DR 79
Loon GR van 92
Lorenzi M 117
Lorenzo AV 11
Loriaux DL 97, 98, 100, 108, 109
Losa M 102, 105, 106, 107, 108, 109, 110, 111, 112, 117, 119, 120
Loughlin JS 88, 92
Love JG 193
Lowe D 31
Lowry PJ 96, 97
Ludt H 29, 38, 59
Lytras N 103, 112

Maas A 43
McCann SM 84, 85, 86, 88, 89, 112, 114
MacCarthy CS 166
McCarthy DM 115
McCraig W 193
McDonald WI 52
McEwan GD 64
McGaffigan P 61, 62, 63
McGarrick G 118
McInnes A 43, 59
Mackinnon IH 21
McLennan JE 191
MacLeod RM 105, 108, 109
McLone DG 196
McNeilly AS 86, 89
Magdelenat H 173
Maie M 64
Malarkey WB 112
Maloof F 79
Maman A 102, 114
Mansfield MJ 88, 92
Mansuy L 172
Marbach P 117
Marcoux RW 27
Marcovitz S 112
Markesbery WR 193
Marshall G 85

Marshall J 31
Marshall JC 86, 89
Marshall LF 43
Martin JB 77
Mason AJ 84, 85, 118
Masten MG 190
Masuda A 97, 103, 106, 109, 114
Mathieu J 190
Matsubara M 129
Matsuda M 35, 43
Matsumura H 5
Matsuo H 74, 84
Maurer IM 21
Mayer DJ 43, 44, 47
Meites J 101
Melmed S 77, 112, 115
Menezes-Ferreira M 80, 81
Merory J 31
Merriam GR 108, 109
Metcalf G 84
Metzger J 173
Meyer CHA 31
Meyer M 31
Michenfelder JD 26
Mickle JP 191
Miller JD 43, 44, 47, 201 ff
Minderhead J 43
Mirimanoff RO 172
Mitang RN 191
Miyakawa M 109
Molitch ME 77, 118
Momma F 40
Monahan JJ 74, 102
Monahan M 74, 75
Monnet F 118
Montgomery WM 61, 62, 63
Morawetz RB 27
More B 114
Morimoto Y 74, 94
Morley JE 77
Morley T 64
Morris DV 91
Moss RL 85
Mougin C 101, 102, 103, 105
Mountjoy CQ 115
Müller EE 77, 90, 111

Müller OA 89, 92, 93, 96, 97, 98, 99, 100, 101, 102, 105, 106, 107, 108, 109, 110, 111, 112, 117, 119, 120
Mulch G 17
Murakami M 64
Murphy WA 102, 109
Mushin J 52
Musuda A 106

Nacimiento AC 49, 59
Nagai M 176
Nahagawa K 129
Nair RMG 74, 84
Nakagami Y 95, 96
Nakagawa T 64
Nakahara M 97, 103, 109
Naor Z 85
Naruse S 186
Negro-Vilar A 114
Neil-Dwyer G 31
Nelson PB 43
Neuenfeldt D 15
Neufeld M 117
Neville H 51, 55, 57
Newlon PG 43
Newman 4
Nicoloff JT 84
Nieschlag E 88
Nikolics K 84, 85, 118
Nillius SJ 88, 92
Nitze M 3
Nornes H 202 ff

Obara N 106, 114
Oblu N 178
Odake J 186
Odell WD 88, 90
Ogata M 5
Ohara H 176
Ohta H 118
Ojeda SR 114
Ojemann RG 27, 62, 63
Oldfield EH 97, 100
Olsen DB 127
Omote K 64
Onofrio BM 189
Oosterom R 97

Oppel F 17
Orci L 113, 115
Ormston BJ 86, 89
Orth DN 97, 98
Overgaad J 43

Padget DH 177
Page LK 11
Page RB 73
Palkovits M 75
Palma L 189
Pandol SJ 103
Parrachi A 80
Partensky C 111
Pástor E 26, 202 ff
Patel YC 75, 79, 93, 113, 115
Patterson J 151, 156
Pearse AGE 113
Pedersen KK 43
Pellerin P 171
Pelletier G 113
Penman E 103
Penny ES 103, 105
Perisutti G 105
Perryman RL 101, 112
Pertuiset B 156, 171, 172, 174, 201 ff
Pertzelan A 127
Petcher TJ 117
Philippon J 172
Philips HS 85
Pickard J 43
Pickard JD 55, 57
Pickardt CR 77, 78, 79, 80, 84
Pictet RL 113
Pieters GFFM 100
Pinchera A 80
Pitts LH 43
Plant TM 85
Pless J 117
Plewe G 117
Poisson M 173
Polak JM 113
Pompili A 171
Poonian MS 74, 102
Powell M 17
Price J 103
Prioleau G 112

Procopio PhT 63
Proulx L 94, 96, 97
Putnam TJ 4

Quigley ME 86

Rabello MM 79, 82
Rabin D 127
Raichle ME 31
Raja IA 195
Ransford AO 64
Ramachandran J 84, 85, 118
Rao K 115
Raptis S 112, 113, 115, 117
Raskin P 115
Raudzens PA 51, 57
Ravazzola M 115
Rebar R 79
Rees JE 31
Rees LH 103
Refetoff S 98
Reichlin S 75, 77, 78, 86, 88, 101, 112, 113, 114, 115, 117, 118
Reid IA 114
Reid SA 191
Reigel DH 43
Reisecker F 64
Renevy F 190
Reschke J 105, 109
Resnik DM 59
Rewcastle NB 189, 190
Rezvani I 79
Rhaney K 186, 189
Richardson PL 31
Richter D 74
Riddervold F 79
Ridgway EC 79
Rifkin RM 102, 114
Rivier C 74, 94, 95, 96, 97
Rivier J 74, 75, 88, 92, 94, 95, 96, 97, 98, 101, 103, 105, 106, 108, 109, 112
Rjosk HK 90, 91, 92
Robert JF 86
Rogol AD 86, 101, 105, 108, 109, 112
Root AW 79
Rose JC 114

Rose LI 78, 114
Rosenbaum TJ 189
Rosenstein J 31, 34
Rosenthal J 112, 113, 115, 117
Ross Russell RW 31
Rossiter VS 59
Rossor MN 115
Roth M 115
Roth SE 113
Rougerie J 4
Rouillar D 114
Rowan JO 31
Roy VM 84
Rubenstein A 115
Rubinstein LJ 196
Rufener C 113
Russell T 16
Rutter WJ 113

Saavedra JM 75
Sachs L 45
Saiki K 64
Saito H 103
Saito S 103
Sakalas R 43
Samson WK 85
Sances A Jr 43
Sandler LM 117
Sandow J 75, 84, 88, 92
Sanes S 190
Sassolas G 111
Satelo-Avila C 191
Scanlon MF 114
Scarff JE 4
Schaff M 112
Schalch DS 84
Schally AV 74, 84, 96, 127
Scheithauer BW 111
Schenk H von 113
Scherer J 51
Schneider V 117
Schopohl J 97, 98, 99, 100, 102, 105, 107, 109, 110, 111, 119, 120
Schulte EH 98
Schulte HM 97, 100
Schürmeyer TH 88
Schusdziarra V 114

Schwartz HG 178
Schweiberer L 112, 117
Schwerdtfeger K 29, 38, 49, 59
Scriba PC 77, 78, 79, 80, 84
Seals DM 59
Sedgwick EM 55, 57
Seeburg PH 84, 85, 118
Seifert H 103
Selhorst JB 51
Selye H 96
Senzoku F 64
Shale DJ 114
Sham C-M 192, 193
Sharbrough FM 26
Sharp PS 106
Shaw NA 28, 30
Sheldon H 97
Shen LP 113
Shibahara S 74, 94
Shibasaki T 97, 103, 106, 109, 114
Shields D 113
Shimatsu A 118
Shizume K 95, 96, 97, 103, 106, 109, 114
Shore RN 79, 82
Siddique T 84
Siesjo BK 27
Siler-Khodr TM 77, 103
Silva IEC 31, 35, 39, 43, 56, 67
Silvestrini F 77, 90, 111
Sinus JK 51
Small R 43
Smals AGH 100
Smith JR 186
Snow MH 114
Snyder PJ 77, 78, 79, 82, 83, 85
Sobieszczyk 105, 117
Sonksen P 91
Sopwith AM 103, 105
Soule EH 189
Spada A 77, 83
Spencer JD 59
Spielman GM 43
Spiess J 74, 94, 95, 96, 97, 105, 106
Squires KC 55
Sriram K 64
Starr A 55, 59

Stalla GK 96, 97, 98, 99, 100, 105, 106, 108, 109, 110, 111, 119, 120
Stears J 51, 55, 57
Sterling FH 79, 82
Stockard JJ 60
Strong AJ 27
Stummvoll HK 79
Suda T 95, 96
Sueiras-Diaz J 96, 102, 109
Sumi SM 192, 193
Sumitomo T 95, 96
Sundt TM Jr 26
Sussman KE 102, 114
Suzuke TA 51
Suzuki M 31
Swanson HS 193
Swanson L 74, 94, 95, 96, 97
Swenson O 191
Swerdloff RS 88, 90
Symon L 26, 27, 31, 35, 39, 40, 43, 54, 57, 60, 201 ff
Szönyi E 84, 85, 118

Taecholarn C 193
Takahashi JS 113
Takaku A 176
Takakuwa K 64
Takano H 64
Takano K 109
Tamaki T 64
Tanjasiri P 114
Tashjian AH Jr 77
Tawana LK 55, 57
Teasdale E 151, 156
Teasdale G 43, 45
Teasdale GM 43, 59
Thibaut A 171
Thomas DGT 64
Thomas DJ 31
Thomas MW 103
Thominet J 105, 112
Thorner MO 97, 101, 105, 106, 109, 112
Thorsen RK 79
Thys CH 86, 89
Tojo K 118
Tomita T 196

Tomori N 95, 96
Torrens MJ 24
Trader S 188
Travaglini P 77, 83
Trojaborg W 26
Trouillas J 111
Tsalikia E 117
Tsuji H 64
Tsushima T 106
Tutty S 16
Tweed WA 43

Überla KK 116
Ugarte N 191
Uhl RR 55
Uitterlinden P 117
Unger RH 113, 114, 115
Unwin RJ 118
Usadel KH 116
Ushiyama T 95, 96
Utiger RD 75, 77, 78, 79, 82, 83

Valcke JC 86, 89
Vale W 74, 75, 86, 92, 94, 95, 96, 97, 101, 103, 105, 106, 109, 112
Valencak E 64
Vance ML 105, 106, 108, 109
Varma SK 97, 101
Vaughan JM 106, 112
Vecht CJ 43
Veilleux R 96
Veldhuis JD 86
Venes JL 196
Verleun T 97
Verschoor L 117
Vijayan E 114
Voigt KH 119
Vouyouklakis D 172
Vugt D van 88
Vulmiere J 4

Wahner HW 79
Wakabayashi I 103, 109
Wakai S 176, 177
Walker AE 177

Walser H 60
Walsh JW 193
Waltz AG 26
Wamatabe M 5
Wang AD 31, 35, 39, 40, 43, 54, 57
Ward DN 74
Ward JD 43
Wardlaw S 88
Warren TG 113
Wass JAH 103
Watanabe K 35, 43
Watters GV 11
Watts C 155
Waxman J 75, 84, 88, 92, 94
Webb CB 105
Webb PJ 64
Webster JD 105
Wehrenberg WB 74, 101, 102, 103, 105
Weightman DR 114
Weiner RI 86, 118
Weinstein ME 59
Weintraub BC 79
Weintraub BD 79, 80, 81
Weissel M 79
Wenzel KW 79
Werder. K von 77, 79, 80, 83, 89, 90, 91, 92, 93, 96, 97, 98, 99, 100, 101, 102, 103, 105, 106, 107, 108, 109, 110, 111, 112, 116, 117, 118, 119, 120
Wheeler M 91
Wher T 79
Wide L 88, 92
Wied D de 73
Wiglesworth FW 186
Wildt L 85, 90, 91, 92
Wilkins RH 195
William B 28, 31
Williams G 79
Williams I 106
Wilson WB 51, 55, 57
Wise BL 195
Witte A 79
Witzmann A 64
Woltman HW 193
Woroch EL 77
Wright JE 55

Yajima F 95, 96
Yamaki T 186
Yamamoto G 106
Yamane T 64
Yamasaki H 64
Yamashita T 64
Yamazaki R 103
Yanetta PJ 63
Yaşargil MG 60, 202 ff
Yen SSC 86

Young HF 43
Young RR 59

Zatz M 113
Zeytin F 74, 101, 102, 103, 105
Zilkha E 31
Zingg HH 113
Zuhlke D 17
Zyznar E 114

Subject Index

Acromegaly 83
  response to
    GnRH 90
    GRH 109, 111
    TRH 83
Addison's disease 96
Adrenocorticotropic Hormone (ADH) 96
  CRH induced secretion 96, 97, 98, 99, 100
Alzheimer disease 115
Amenorrhea
  hypothalamic 91
  response to GnRH 89
Anesthesia
  recording of
    SSEP 38
    VEP 55
  systemic arterial pressure (SAP) 36
Aneurysms, saccular
  arterial hypotension 205
  brain swelling 203
  hematoma 203
  intraoperative
    management 201
    rupture 208, 209, 210
  management of the spasm of the ICA 208
  mean arterial pressure monitoring 204
  misapplication of clip 42
  place of CSF drainage 201
    advantages 201
    disadvantages 202
  problem of dural tension 202
  pulmonary artery pressure monitoring 204
  surgery and SSEP 35, 37
  technique of
    coating 207
    occlusion of the neck 207
  temporary clipping 205
  use of
    bipolar coagulation 206
    cardiac arrest and hypothermia 209
    loupes 204
    microscope 204
    self retaining retraction 204
Angiography
  in sphenoidal ridge meningioma 148, 149, 150, 164
Anorexia nervosa
  response to GnRH 90
Antidiuretic Hormone (ADH) 76, 83
Aqueduct, stricture of 12, 15
  treatment by endoscopic Seldinger technique 15
Arnold-Chiari malformation 196

Bipolar coagulation
  in aneurysm surgery 206
Brain Stem Auditory Evoked Potentials (BAEP)
  see Evoked Potentials
  in
    acoustic neuroma surgery 61
    head injury 58
    posterior fossa surgery 59
Bulb
  incandescent 3, 5, 8, 9
  quartz halogen 9

Calmodulin 77
Central Conduction Time (CCT) 28, 32
  and
    anesthesia 39, 42
    brain retraction 39
    misapplication of clip 42
  as prognostic indicator 35
  clinical significance 34
  definition 30
  during temporary vascular occlusion 40
  in
    head injury 44, 50
    subarachnoid hemorrhage 31
  interhemispheric difference 29, 32, 33, 35, 44

CCT, normal values 29
　relationship with CBF 34, 36
Cerebello pontine angle
　endoscopic expolation 17
Cerebral Blood Flow (CBF)
　disruption of ionic homeostasis 27
　in subarachnoid hemorrhage 31
　relationship with CCT 34, 36
Cerebro Spinal Fluid (CSF)
　artificial CSF solution 22
　as light absorber 9
　CSF drainage in management of ruptured saccular aneurysms 201–202
　physiology 11
　postoperative leakage 4
Choroid plexus coagulation 12
　indications 13
　results 14
　technique 12
CO_2 laser
　in surgery of clinoidal meningioma 167
Corticotropin Releasing Hormone (CRH) 74, 76, 94, 119
　distribution 95
　in diagnosis of
　　adrenal failure 98
　　Cushing's syndrome 100
　measurement 95
　pathophysiology 97
　role in the stress response 96
　structure 94
　test 75
　therapeutic significance 101
Cushing's
　disease 83, 97, 98, 101
　syndrome 96, 100, 101
Cystoscope 3, 4
Cysts 176
　arachnoid 177, 195
　　intradural 195
　dermoid and epidermoid 177, 180, 184
　　clinical features 179, 180
　　embriology 177
　　intracranial 178
　　intramedullary 183
　　surgical management 183
　glial 177
　neurenteric 177, 188
　　clinical features 186
　　embriology 185

　　radiology 187
　　surgical management 187
　teratomatous 177, 188
　　clinical features 189, 190
　　differentiating from neurenteric cysts 189
　　embriology 188

Dermoid sinus 178, 179, 181, 182, 183, 184
Diabetes insipidus 121
Diastematomyelia 180, 182, 186, 190, 191
Dopamine 74, 118, 121
　control of
　　GnRH release 86, 118
　　PRL secretion 118

Electrocochleogram
　in acoustic neuroma surgery 61, 63, 64
Electrodes
　in recording
　　SSEP 28–29
　　VEP 52
Electroencephalogram 26, 27, 28, 60
　in sphenoidal ridge meningioma 146
Endoneurosurgery
　see Endoscopic Intracranial Surgery
Endoscope 3, 4, 5, 10, 15, 16, 19, 22
　applications 11
　light sources 7, 8
　　colour balance 8, 20
　　concentration 8
　　heat output 9
　Nitze system 3, 4, 5, 6, 8
　small flexible system 19
　　advantages 20
　　disadvantages 20
　solid rod lens system 5, 6, 20
Endoscopic Intracranial Surgery
　endoscopic photography 16–17
　exploration of
　　cerebellopontine angle 17
　　cisterna magna 21
　historical introduction 3
　intraventricular tumor biopsy 15–16
　sterilising technique of the endoscope 20–21
　transsphenoidal surgery 17
　treatment of
　　colloid cysts 17
　　hydrocephalus 11–15
　　spinal lumbar pain 22
　　trigeminal neuralgia 21

Subject Index

Epilepsy
   in sphenoidal ridge meningioma 140
Evoked Potentials
   auditory (BAEP) 26
      head injury 58
      acoustic neuroma surgery 61
      posterior fossa surgery 59
   central conduction time (CCT) 28
   somatosensory 26
      aneurysm surgery 37
         anesthesia 38
         temporary vascular occlusion 40
      head injury management 43
         posttraumatic changes 49
      spinal cord monitoring 64
      stimulation 29
      subarachnoid hemorrhage 30
      technique of recording 28–30
      vascular disease of the brain 26
   variability of amplitude 27
   visual 26, 51
      anesthesia 55
      changes-grading classification 56
      during surgery 55, 57
      methodology 52
      normal values 54
      posttraumatic 51

Fertile eunuch syndrome 88
Follicle Stimulating Hormone (FSH) 85
Foramen of Monroe, cysts
   endoscopic treatment 17
Forster Kennedy syndrome 142

Glasgow
   Coma Scale (GCS) 45, 46, 50, 59
   Outcome Scale 46, 50
GnRH—Associated Peptide (GAP) 84, 118
Gonadotropin Releasing Hormone 74, 76, 119
   desensitization of GnRH receptors 88
   distribution 85
   physiological role 85
   in diagnosis of
      acromegaly 90
      gonadal disorders 89
      hyperprolactinemic disorders 90
   stimulation of gonadotropin secretion 86
   structure 84
   superactive agonists 92
   therapeutic significance 90

Graves disease 79
Greenberg scheme 45, 48
Growth Hormone (GH)
   GRH stimulated secretion 104, 106, 108
   TRH stimulated secretion 83
Growth Hormone Releasing Hormone (GRH) 74, 76, 119
   acromegaly 109
   biological activity 101, 105
   diagnosis and treatment of pituitary dwarfism 109
   distribution 103
   evaluation of anterior pituitary function 107
   GRH producing tumor 112
   measurement 103
   monoclonal GRH antibodies 105
   stimulation of GH release 104, 106, 108
   structure 101

Head injury
   posttraumatic changes 49
   recording of
      BAEP 59
      SSEP 43
Hensen's node 189
Heubner's artery 43
Homeostasis, ionic
   and falling CBF 27
Horner's syndrome 186
Hunt and Hess, scale of 31
Hydrocephalus
   choroid plexus coagulation 12, 14
      indications 13
      results 13
   classification 14
   extirpation of the choroid plexus 4
   ICP monitoring 13
   pathophysiology 11
   photographs of the cerebral ventricles 4
Hyperthyroidism 78, 79, 80, 82
Hypogonadism 121
   hypogonadotropic 88, 92
   hypothalamic 88, 92
   primary 89
   secondary 89
   tertiary 90
   treatment with GnRH 92
Hyposomatotropism 121
Hypotension, controlled
   in aneurysm surgery 205

Hypothalamus 73, 103
  causes of hormone hypersecretion 121
  hormones and peptides 76
  lesions 81
  therapy of hormone deficiency 121
Hypothyroidism 82, 121
  hypothalamic 80, 81
  primary 80, 83
    TSH response to TRH 80
  secondary 80
  tertiary 78

Infarction, cerebral
  pathophysiology 27
Insulin-Induced Hypoglycemia Test (IHT) 106, 108, 110
Interhemispheric difference (IHD)
  in SSEP 29, 32, 33, 35, 44
Intra Cranial Pressure (ICP)
  monitoring of 13

Kallmann's syndrome 88, 89, 90, 93
Kovalesky, canal of 187

Langerhans, islets of 75
Light
  cool 7
  fluorescent 8, 9
  monochromatic 8
  Storz light box 9
Lilliquist's membrane 203
Lipoma 177
  clinical features 193
  differentiating from dysraphic lipomatous malformation 193
  embriology 192
  intramedullary 193, 194
Lipomeningocele 191
Luteinizing Hormone (LH) 85
  response to GnRH 86, 87, 89

Magnetic Resonance Imaging (MRI)
  in sphenoidal ridge meningioma 147
Mannitol 157
Meningioma of the sphenoidal ridge
  clinoidal 145, 146, 148, 149, 154
    surgical approach 158–168
      craniotomy 159, 160
      control of insertion of the tumor 167
      dural closure 167
      dural opening 160
      removal of
        tumor 165
        capsule 166
      scalp incision 159, 160
      uncapping the tumor 163
  en plaque 139, 140, 141, 144, 151, 172
    surgical approach 170
  pterional 143, 144, 147, 149, 150, 151
    surgical approach 159, 168
      dural
        closure 170
        opening 169
Meningocele 178, 180, 196
Minisomatostatin 117
Mirror
  dichroic 9
  parabolic 9
Myelocystocele 178
Myelomeningocele 178, 180, 182, 183, 186, 191, 195, 196

Nd:Yag laser
  in surgery of clinoidal meningioma 167
Nelson's syndrome 96
Neuroblastoma 176
Neurohormones, hypothalamic hypophysiotropic 73
  see CRH, GnRH, GRH, SRIF, and TRH
  biological significance 75
Neuroschisis 177, 178, 186
Notochord 186

Oxytocin (OT) 76

Papaverine 208, 209
Parkinson's disease 115, 119
Pearson's contingence effect 47, 49
Peptides
  endogenous opioid 86
  hypophysiotropic 73, 74, 76
  of the posterior lobe 73, 76
Phakomatoses 176
Photography, endoscopic 16–17
Pituitary
  (gland), anterior 73
    control of function 74
  portal system 73, 74, 77, 85, 113, 118
    tuberoinfundibular neurons (TIDA) 118
  tumor
    response to GnRH 89
    TSH response to TRH 80, 81

Subject Index

Prolactin (PRL)
    disorders of secretion 82
    factor
        inhibiting 74, 76, 117, 118
        releasing 76, 77, 118
    response to
        GRH 110
        TRH 82
Prolactinoma 83, 119
Proopiomelanocortin (POMC) 96, 97, 110
Pseudopuberty, precocious 88
Psychiatric disorders
    in sphenoidal ridge meningioma 141
Pterional meningioma
    see Sphenoidal ridge meningioma
Puberty, precocious 88, 92

Reflection, total internal 7
Refraction 7
Refractive index 5, 6, 7

Sarcoidosis
    endoscopic intraventricular biopsy 16
Seldinger wire 15
Sodium nitroprusside 158
Somato Sensory Evoked Potentials (SSEP)
    see Evoked Potentials
    anesthesia 38
    aneurysm surgery 37
    central conduction time 28
    head injury management 43
        posttraumatic changes 49
    spinal cord monitoring 64
    stimulation 29
    subarachnoid hemorrhage 30
    technique of recording 28–30
    vascular disease of the brain 26
Somatomedin C 102, 109
Somatostatin 74, 75, 76, 102, 110
    clinical use 115
    distribution 113
    GRH induced suppression 106
    measurement 114
    minisomatostatin 117
    pathophysiology 115
    physiological role 114
Somatotropin—Release Inhibitory Factor (SRIF)
    see Somatostatin
Sphenoidal bone
    ala magna 138

    ala parva 138
Sphenoidal ridge meningioma 137
    angiography 148, 149, 150, 164
    classification of tumors 139
    clinical examination 141, 142
    CT scan 143, 145, 152
    description of sphenoidal ridge 138
    EEG 146
    en plaque 139, 140, 141, 144
    first symptoms 139
    Gamma scan 146
    incidence 139
    MRI 147
    surgical treatment
        anesthesia 157
        considerations 151
        indications 155
        positioning 158
        postoperative care 171
        preoperative care 155
        results 171
        technique 159
Spina bifida 176, 177, 186, 189, 193
    aperta 178, 180
    cystica 178
    occulta 195
Spinal cord congenital malformations, in children
    lipomyelomeningocele 176
    venous angiomas 176
Spinal cord congenital tumors, in children
    clinical features 176, 179, 180, 186, 189
    cysts 176
        arachnoid 177, 195
        dermoid 177, 180
        epidermoid 177, 180
        neurenteric 185, 187, 188
    definition 176
    embriology 176, 177, 185, 188
    incidence 176
    lipoma 177
    malignant tumors 196
    neuroblastoma 176
    radiology 176, 182, 187
    surgical management 183, 187
Sterilisation
    of endoscopes 21
Subarachnoid hemorrhage
    measurement of SSEP 30
Swann Ganz catheter 204

Telescope
   approach to the ventricle 10
   "fore oblique" 4, 17
   in
      infants 10
      transsphenoidal surgery 17
Teratoma, intraspinal
   clinical features 191
   embriology 191
   sacrococcygeal 191, 193
   trigerminal 189
Thyroid Stimulating Hormone (TSH) 78
   αTSH 80
   βTSH 80
   enzyme-immunoassays 80
   response to
      somatostatin 114
      TRH 79
   thyrotoxicosis 79
Thyrotropin Releasing Hormone (TRH) 74, 76, 119
   biosynthesis 77
   clinical utilization 78
   distribution 75
   function 77
   in diagnosis of
      acromegaly 83
      disorders of PRL secretion 82
      hypothalamic-pituitary disease 80
      thyroid disorders 78
   pathophysiology 78
   structure 75
   test 75, 79
   therapeutic significance 84
Tuberoinfundibular neurons (TIDA) 118

Vasoactive Intestinal Polypeptide (VIP) 77, 118
   control of PRL secretion 118
Ventriculoscope 5, 20
   Hopkins solid rod system 20
Ventriculostomy 4, 15
Video film, endoscopic 16
Visual Evoked Response
   see Evoked Potential
   anesthesia 55
   changes-grading classification 56
   during surgery 55, 57
   methodology 52
   normal values 54
   posttraumatic 51

Xenon
   arc 8, 9, 16
   clearance technique 31

Advances and Technical Standards in Neurosurgery

Volume 1

1974. 96 figures. XI, 210 pages.
Cloth DM 102,—, öS 714,—
ISBN 3-211-81218-0

Contents:

Advances: N. Lundberg, Å. Kjällquist, G. Kullberg, U. Pontén, and G. Sundbärg: Non-operative Management of Intracranial Hypertension. — J. Philippon and D. Ancri: Chronic Adult Hydrocephalus. — H. Powiertowski: Surgery of Craniostenosis in Advanced Cases. A Method of Extensive Subperiosteal Resection of the Vault and Base of the Skull Followed by Bone Regeneration. — E. Zander and R. Campiche: Extra-Dural Hematoma.

Technical Standards: B. Pertuiset: Supratentorial Craniotomy. — B. Guidetti: Removal of Extramedullary Benign Spinal Cord Tumours.

Volume 2

1975. 150 partly coloured figures. XI, 217 pages.
Cloth DM 113,—, öS 790,—
ISBN 3-211-81293-8

Contents:

Advances: J. Gawler, J. W. D. Bull, G. du Boulay, and J. Marshall: Computerized Axial Tomography with the EMI-Scanner. — M. Samii: Modern Aspects of Peripheral and Cranial Nerve Surgery. — A. Rey, J. Cophignon, Cl. Thurel, and J. B. Thiebaut: Treatment of Traumatic Cavernous Fistulas.

Technical Standards: M. G. Yaşargil, J. L. Fox, and M. W. Ray: The Operative Approach to Aneurysms of the Anterior Communicating Artery. — Valentine Logue: Parasagittal Meningiomas. — J. Siegfried and M. Vosmansky: Technique of the Controlled Thermocoagulation of Trigeminal Ganglion and Spinal Roots.

Volume 3

1976. 77 figures. XI, 154 pages.
Cloth DM 92,—, öS 644,—
ISBN-3-211-81381-0

Contents:

Advances: G. Guiot and P. Derome: Surgical Problems of Pituitary Adenomas. — H. Troupp: The Management of Intracranial Arterial Aneurysms in the Acute Stage. — Y. Yonekawa and M. G. Yaşargil: Extra-Intracranial Arterial Anastomosis: Clinical and Technical Aspects. Results.

Technical Standards: W. Luyendijk: The Operative Approach to the Posterior Fossa. — J. Brihaye: Neurosurgical Approaches to Orbital Tumours. — R. Lorenz: Methods of Percutaneous Spino-Thalamic Tract Section.

Advances and Technical Standards in Neurosurgery

Volume 4
1977. 66 partly coloured figures. XI, 154 pages.
Cloth DM 92,—, öS 644,—
ISBN 3-211-81423-X

Contents:

Advances: N. A. Lassen and D. H. Ingvar: Clinical Relevance of Cerebral Blood Flow Measurements. — G. W. Taylor and J. S. P. Lumley: Extra-Cranial Surgery for Cerebrovascular Disease. — J. Rétif: Intrathecal Injection of a Neurolytic Solution for the Relief of Intractable Pain.

Technical Standards: L. Symon: Olfactory Groove and Suprasellar Meningiomas. — M. G. Yaşargil, R. D. Smith, and J. C. Gasser: Microsurgical Approach to Acoustic Neurinomas. — G. Debrun, P. Lacour, and J. P. Caron: Balloon Arterial Catheter Techniques in the Treatment of Arterial Intracranial Diseases.

Volume 5
1978. 78 figures. XII, 224 pages.
Cloth DM 123,—, öS 860,—
ISBN 3-211-81441-8

Contents:

Advances: A. M. Landolt: Progress in Pituitary Adenoma Biology. Results of Research and Clinical Applications. — J. Hildebrand and J. Brihaye: Chemotherapy of Brain Tumours. — S. Mingrino: Supratentorial Arteriovenous Malformations of the Brain.

Technical Standards: J. Hankinson: The Surgical Treatment of Syringomyelia. — F. Loew and W. Caspar: Surgical Approach to Lumbar Disc Herniations. — B. Pertuiset, D. Fohanno, and O. Lyon-Caen: Recurrent Instability of the Cervical Spine With Neurological Implications— Treatment by Anterior Spinal Fusion.

Volume 6
1979. 79 figures. XI, 191 pages.
Cloth DM 113, —, öS 790,—
ISBN 3-211-81518-X

Contents:

Advances: E.-O. Backlund: Stereotactic Radiosurgery in Intracranial Tumors and Vascular Malformations. — J. Klastersky, L. Kahan-Coppens, and J. Brihaye: Infection in Neurosurgery. — C. Gros: Spasticity-Clinical Classification and Surgical Treatment.

Technical Standards: P. J. Derome and G. Guiot in co-operation with B. Georges, M. Porta, A. Visot, and S. Balagura: Surgical Approaches to the Sphenoidal and Clival Areas. — R. Braakman: Cervical Spondylotic Myelopathy. — F. Isamat: Tumours of the Posterior Part of the Third Ventricle: Neurosurgical Criteria.

Volume 7
1980. 147 figures. XI, 247 pages.
Cloth DM 125,—, öS 875,—
ISBN 3-211-81592-9

Contents:

Advances: M. G. Yaşargil, R. W. Mortara, and M. Curcic: Meningiomas of Basal Posterior Cranial Fossa.

Technical Standards: A. M. Landolt and P. Strebel: Technique of Transsphenoidal Operation for Pituitary Adenomas. — Surgical Treatment of Facial Nerve Paralysis; Longterm Results: H. Millesi: Extratemporal Surgery of the Facial Nerve — Palliative Surgery. S. Mingrino: Intracranial Surgical Repair of the Facial Nerve. U. Fisch: Management of Intratemporal Facial Palsy.

Advances and Technical Standards in Neurosurgery

Volume 8

1981. 135 partly coloured figures. XII, 328 pages.
Cloth DM 157,—, öS 1100,—
ISBN 3-211-81665-8

Contents:

Advances: E. de Divitiis, R. Spaziante, and L. Stella: Empty Sella and Benign Intrasellar Cysts. — B. Pertuiset, D. Ancri, and A. Lienhart: Profound Arterial Hypotension (MAP $\leq$ 50 mm Hg) Induced with Neuroleptanalgesia and Sodium Nitroprusside (Series of 531 Cases). Reference to Vascular Autoregulation Mechanism and Surgery of Vascular Malformations of the Brain. — F. Gullotta: Morphological and Biological Basis for the Classification of Brain Tumors. With a Comment on the WHO-Classification 1979.

Technical Standards: E. Pásztor: Surgical Treatment of Spondylotic Vertebral Artery Compression. — P. Harris, I. T. Jackson, and J. C. McGregor: Reconstructive Surgery of the Head. — A. N. Konovalov: Operative Management of Craniopharyngiomas.

Volume 9

1982. 88 figures. XI, 177 pages.
Cloth DM 102,—, öS 714,—
ISBN 3-211-81718-2

Contents:

Advances: K. Faulhauer: The Overdrained Hydrocephalus. Clinical Manifestations and Management. — A. P. Romodanov and V. I. Shcheglov: Intravascular Occlusion of Saccular Aneurysms of the Cerebral Arteries by Means of a Detachable Balloon Catheter. — H. Spiess: Advances in Computerized Tomography.

Technical Standards: L. Symon: Surgical Approaches to the Tentorial Hiatus. — F. Loew: Management of Chronic Subdural Haematomas and Hygromas. — B. Williams: Subdural Empyema.

Volume 10

1983. 70 figures (1 in color). XI, 231 pages.
Cloth DM 123,—, öS 860,—
ISBN 3-211-81750-6

Contents:

Advances: R. J. S. Wise, G. L. Lenzi, and R. S. J. Frackowiak: Applications of Positron Emission Tomography to Neurosurgery. — J. Siegfried and T. Hood: Current Status of Functional Neurosurgery. — B. Pertuiset, D. Ancri, J. P. Sichez, M. Chauvin, E. Guilly, J. Metzger, D. Gardeur, and J. Y. Basset: Radical Surgery in Cerebral AVM — Tactical Procedures Based upon Hemodynamic Factors.

Technical Standards: M. Sindou and A. Goutelle. Surgical Posterior Rhizotomies for the Treatment of Pain. — A. Kumar and U. Fisch: The Infratemporal Fossa Approach for Lesions of the Skull Base.

Advances and Technical Standards in Neurosurgery

Volume 11

1984. 1 portrait. 80 figures. XII, 248 pages.
Cloth DM 125,—, öS 875,—
ISBN 3-211-81806-5

Contents:

Hugo Krayenbühl — An Appreciation (By M. G. Yaşargil)

Advances: G. M. Bydder: Nuclear Magnetic Resonance Imaging of the Central Nervous System — G. Huber and U. Piepgras: Update and Trends in Venous (VDSA) and Arterial (ADSA) Digital Subtraction Angiography in Neuroradiology.

Technical Standards: M. G. Yaşargil, L. Symon, and P. J. Teddy: Arteriovenous Malformations of the Spinal Cord — C. Lapras, R. Deruty, and Ph. Bret: Tumors of the Lateral Ventricles — F. Loew, B. Pertuiset, E. E. Chaumier, and H. Jaksche: Traumatic, Spontaneous and Postoperative CSF Rhinorrhea.

Volume 12

1985. 49 partly colored figures. XI, 186 pages.
Cloth DM 114,—, öS 798,—
ISBN 3-211-81877-4

Contents:

Advances: Valerie Walker and J. D. Pickard: Prostaglandins. Thromboxane, Leukotrienes and the Cerebral Circulation in Health and Disease.

Technical Standards: M. G. Yaşargil, P. J. Teddy, and P. Roth: Selective Amygdalo-Hippocampectomy. Operative Anatomy and Surgical Technique — E. Pásztor: Transoral Approach for Epidural Craniocervical Pathological Processes.

Volume 13

1986. 77 partly colored figures. IX, 179 pages.
Cloth DM 114,—, öS 798,—
ISBN 3-211-81885-5

Contents:

Advances: J. M. Tew Jr., and W. D. Tobler: Present Status of Lasers in Neurosurgery.

Technical Standards: H. G. Wieser: Selective Amygdalohippocampectomy: Indications, Investigative Technique and Results — F. Epstein: Spinal Cord Astrocytomas of Childhood.

Springer-Verlag Wien New York

Mölkerbastei 5, A-1011 Wien
175 Fifth Avenue, New York, NY 10010, USA
Heidelberger Platz 3, D-1000 Berlin 33
37-3, Hongo 3-chome, Bunkyo-ku, Tokyo 113, Japan

MIX
Papier aus verantwortungsvollen Quellen
Paper from responsible sources
FSC® C105338

If you have any concerns about our products,
you can contact us on
ProductSafety@springernature.com

In case Publisher is established outside the EU,
the EU authorized representative is:
**Springer Nature Customer Service Center GmbH
Europaplatz 3, 69115 Heidelberg, Germany**

Printed by Libri Plureos GmbH
in Hamburg, Germany